CHEATING DE

"Through nuanced profiles of patients, doctors, researchers, and activists, Hirsch persuasively illustrates an epidemic that is at odds with modern society at almost every level . . . Hirsch has written the book that people who care about diabetes have been waiting for. If it spurs more of us to work for a cure, so much the better."
— *Washington Post*

"Powerful."
— *St. Louis Post Dispatch*

"Hirsch has an insider's candor speaking about life with diabetes, the sensitivity of the parent of a child with a chronic illness, and skill of a good journalist reporting on the medical, social, economic, and scientific details of what was once called 'the wasting disease.'"
— *Booklist,* **starred review**

"Written with both craft and passion . . . this book makes its case with skillful writing and emotional impact."
— *Library Journal*

"Skillfully combin[es] journalistic expertise with his personal story . . . this is an informative and moving analysis."
— *Publishers Weekly*

"A character-driven account, written with barely controlled anger, of what diabetes is, what it is like to be diabetic, and how and why the medical community is failing to deal effectively with this widespread and as-yet-incurable condition . . . A provocative amalgam of staunch advocacy, personal experience, and investigative reporting."
— *Kirkus Reviews,* **starred review**

CHEATING DESTINY

BOOKS BY JAMES S. HIRSCH

HURRICANE
The Miraculous Journey
of Rubin Carter

RIOT AND REMEMBRANCE
The Tulsa Race War
and Its Legacy

TWO SOULS INDIVISIBLE
The Friendship That Saved
Two POWs in Vietnam

CHEATING DESTINY
Living with Diabetes,
America's Biggest Epidemic

CHEATING DESTINY

LIVING WITH DIABETES

James S. Hirsch

A MARINER BOOK
HOUGHTON MIFFLIN COMPANY
BOSTON • NEW YORK

First Mariner Books edition 2007

www.houghtonmifflinbooks.com

The Library of Congress has cataloged the hardcover edition as follows:
Hirsch, James S.
Cheating destiny: living with diabetes, America's biggest
epidemic / James S. Hirsch.
p.   cm.
Includes bibliographical references and index.
ISBN-13: 978-0-618-51461-8
ISBN-10: 0-618-51461-9
1. Diabetes — Popular works. I. Title.
RC660.4.H57 2006
616.4'62 — dc22   2006011239

ISBN-13: 978-0-618-91899-7 (pbk.)
ISBN-10: 0-618-91899-X (pbk.)

Book design by Melissa Lotfy

PRINTED IN THE UNITED STATES OF AMERICA

MP 10 9 8 7 6 5 4 3 2 1

To Lynn Friedman and Irl Hirsch,
for looking after their younger brother.

To the Carter Center for Children,
for going beyond the call of duty.

Contents

CHEATING DESTINY

Prologue

MY SON IS THIRSTY.

For most parents that sounds rather harmless, one of many needs that any three-year-old has. At first, Garrett's request seems particularly innocuous. He has always preferred drinking milk or juice to eating. His constant running and playing would dehydrate any child, and he shows no sign of illness. But when he looks up at me and says, "Daddy, I'm thirsty," repeatedly over several days, I wonder if the damage has already been done.

Soon Garrett's complaint becomes more urgent, and the water flows right through his small body, causing frequent trips to the bathroom. I try to break the cycle, telling him he's had enough to drink and hoping that his craving will go away. But still he drinks. He is potty trained, and he does all he can to hold the tidal wave of water coursing through him. On several occasions, he groans in his sleep to avert wetting his bed. I rush into his room, hustle him into the bathroom, and yank down his pajamas just in time. *Splash!* The urine rushes out like water from a fire hose. A couple of times the force is too great, and he pees in his bed.

"That's okay, buddy," I tell him. "We just can't drink so much water at bedtime."

I am hoping for something, anything, whatever it takes to diminish his longing. But the water has become his lifeline. I watch him hold the glass in both hands, lift it carefully to his mouth, tilt it, and swallow again and again. I had once been thirsty like that. It was many years ago, but I remember it well.

* * *

1

In medical-speak, the word is *polydipsia* — abnormal thirst. It's an early symptom of diabetes: your body, suffering from elevated blood sugar, pees out the excess glucose and triggers the demand for more water. As the disease progresses, the body burns its own fat for energy, leading to a similar cycle of insatiable hunger followed by rapid weight loss. These are signs of type 1 diabetes, which is usually diagnosed in juveniles and treated with insulin.

I was diagnosed with type 1 at fifteen, and like every parent with diabetes, I scrutinize my kids — we also have a six-year-old daughter, Amanda — every day for symptoms, real or imagined. Hunger. Thirst. Fatigue. Weight loss. Emergency trips to the bathroom. Cuts that heal slowly. Crankiness. Unusual cravings. Any aberrant behavior could be a sign, however tenuous, of disorder in a child's finely tuned metabolic system. As a diabetic, I learned early on that the price of health is eternal vigilance, but as a parent, the price of devotion is chronic paranoia.

Garrett, at this stage, displays no other symptoms. He hasn't lost weight, increased his appetite, or complained of tiredness. On the contrary, he cannot look any better or behave any more vibrantly, a high-spirited little boy with tousled sandy hair and limpid brown eyes. His preschool teachers call him "Smiley" because he's always laughing. He loves sports — running, tackling, kicking a soccer ball — and is already hitting live pitching in our driveway. He's strong-willed — a nice way of saying he's stubborn. One night we heard a loud thump in his room. He had climbed out of his crib and crashed to the hardwood floor. We figured he had learned his lesson and put him back in the crib. Minutes later, *thump!* He had repeated his escape to the floor. At eighteen months, he was out of the crib. His pediatrician says that, pound for pound, he's the strongest patient she has.

But all the signs of health are now misleading. A nagging cold has slowed him down, and despite my coaxing, his desire for water continues to be strong. Events take an eerie turn one day in September 2004 when I interview Jeff Hitchcock in Boston for this book. Hitchcock's daughter, Marissa, was diagnosed with diabetes in 1989. Finding little medical information, he started a Web site about diabetic children from his home in Hamilton, Ohio. The site was so wildly successful — 250,000 hits per day from 149 countries and a raft of advertisers — that Hitchcock quit his engineering job and now organizes conferences and programs as well. He is revered by parents who feel neglected by health care professionals and find his site informative and comforting. When I meet Hitchcock, I understand his appeal. A lean man with graying hair, glasses, and a soothing demeanor, he speaks optimistically about the day that improved therapies will eliminate diabetic complications. But he also directs stinging criticism at the medical field. This combination of hope and frustration resonates with any diabetic. Asked what the success of his site reveals about diabetic care in America, he says, "It stinks." And what should parents do if their child is receiving poor care? "Fire the doctor," he says. "That doesn't happen nearly enough."

That night, Garrett complains during his bath that his feet and legs hurt, and he again drinks several cups of water. He goes to bed but wakes up around midnight, saying he doesn't feel well. He is thirsty, but I tell him he just had a drink a few hours earlier. He again goes to the bathroom and, sniffling and achy, climbs into our bed. I look at him in his baseball pajamas, pale and uncomfortable. Until now, I have not mentioned anything to Sheryl, my wife, about my fears. She knows about the genetic risks of diabetes,* but when I tell her that I'm going to test Garrett's blood sugar, she's surprised.

The glucose meter measures blood sugar in milligrams per deciliter. The normal fasting range is less than 100 mg/dl. Between 100 and 125 represents "impaired glucose fasting," so 125 is the magic number — anything higher is a sign of diabetes. I take my lancet and quickly poke Garrett's finger for a drop of blood. He is too groggy to complain or even notice. The blood comes out easily in a thick, gooey drop. The older meters took thirty seconds or even a minute to read the value, but the newer meters — mine is a OneTouch Ultra from Lifescan — has a five-second countdown. I place the drop of blood on the test strip and prepare myself. But the whole thing is anticlimactic. I already know the result.

5-4-3-2-1.

---

* According to C. Ronald Kahn, president of the Joslin Diabetes Center, a child whose father has type 1 diabetes has a 7 percent chance of developing the disease. If the mother has type 1, the child has a 2 percent chance. Otherwise, the child has between a 0.3 and 0.4 percent chance of getting it.

The machine reads HI.

I have never seen such a reading and am momentarily confused. HI? Why the hell is the machine suddenly greeting me? No, no, no. Not HI as in HELLO, HI as in HIGH. As in: real HIGH. As in: your entire life has just changed. I curse Lifescan under my breath. With the millions of dollars it makes from diabetics, you'd think it could afford to put two more letters on an elevated glucose reading instead of subjecting us to this incongruously cheerful, heartbreaking welcome.

"He's high," I tell Sheryl. "I think we'll have to take him to the hospital."

She grabs Garrett and hugs him, and will soon dry her tears with the bloodstained tissue that I used to wipe his finger. I get him a water bottle and apologize for not letting him drink more. I have tried to deny his body's downward spiral, to will it back to health by limiting his fluids. But my son is thirsty. "Here, buddy, drink this," I say. "Drink as much as you want."

I call my brother, Irl, who's had diabetes since he was a child and now, as an endocrinologist, runs a large diabetes clinic in Seattle for the University of Washington. He asks if Garrett has ketones, a fatty acid burned by the body of an uncontrolled diabetic that spills out in the urine. Almost 1,900 diabetics die each year from ketoacidosis, and a small child, once ketonic, can become fatally ill. A simple urine or home blood test can determine the presence of ketones. I tell my brother that I don't know yet, but I will keep him posted.

Sheryl calls the answering service for Garrett's physician; with no one on call, the operator bounces us to another practice. A doctor finally gets on the line and tells us to take Garrett to Children's Hospital Boston, about a half-hour drive from our house in Needham. I had been across the street from Children's earlier that day when I spoke with Hitchcock at the Joslin Diabetes Center. We talked about his Web site, Children with Diabetes. Now I will return that night with my son in the back seat — a child with diabetes.

Our daughter is sleeping, so Sheryl stays home while I take Garrett. When he sees her packing his overnight bag, he's excited. "Are we going to a pajama party?" he asks.

"No, buddy," I say. "We aren't going to a pajama party."

4

I tell him we're going to the hospital without explaining why. He has never really been sick and hadn't been in a hospital since he was born. We load up the Honda Pilot, and Sheryl kisses Garrett and me good-bye. We briefly hop on the highway, then take Route 9 toward Boston. The road, normally chaos, is now dark and quiet, with streetlamps splashing islands of light on the pavement. The city is at peace. Soft music drifts through the car. Still in his baseball pajamas, Garrett looks calmly out the window, probably thinking he's going on an adventure, a late-night ride to the hospital — or "HOTH-i-bal," as he calls it, with a slight, endearing lisp — something he can tell his friends about the next day.

We arrive in the Emergency Room at 2 A.M., Garrett in my arms. The place is empty. A man at the information desk points me to a woman in a cubicle who will handle Garrett's admission. We sit down.

"Do you have insurance?" She does not look up.

I guess if you enter a hospital in the dead of night with your son bleeding from an open wound, choking, or screaming, someone will first ask if you need medical assistance. Otherwise, reimbursement takes priority.

"Yes, we have insurance."

We are sent to a room where a nurse takes some information and Garrett is weighed — thirty-three pounds. Then we move to another room. Garrett sits on my lap, and I assure him that everything is going to be okay. He doesn't ask any questions as hospital staff members drift in and out with glucose machines, needles, tubes, and other devices. It appears that we are the only customers on the floor, and a calm settles in. But that soon ends.

The nurses' first task is extracting Garrett's blood. They initially use a lancet to get a drop from his finger, which they test in a meter. But now they need tubes of blood, not just drops, so they have to draw it from his vein. I hold Garrett on the table while one of the nurses positions the needle above his arm. As she drives it through the flesh, Garrett lets out a scream unlike any I have ever heard. Shocked by the attack, he yells, squirms, grits his teeth, and howls some more. "Daddy, that hurts me! Daddy, that hurts me!"

"I know, buddy, but they're almost done. They're almost done."

But they aren't. The nurse pulls back the plunger but draws no blood. She tries rotating the needle, increasing the pain and getting more resistance from Garrett. But she has missed the target, so she removes the needle and plunges it again into his arm. Garrett tries to escape, but I hold him down. Tears are rolling down his reddened face as he yells again, "That hurts me! That hurts me!" The nurse explains that it's often hard to hit a vein in a young child and says she will try the back of his hand instead — which, for my money, does not seem particularly promising. I've had blood drawn well over a hundred times, always from the arm. The puncture stings, but at least the arm has some cushion, while the hand is as hard and unforgiving as parched earth.

The needle lunges into Garrett's hand, and this time he closes his eyes and cries even louder. Still no blood. The nurse again maneuvers the needle, fruitlessly inflicting more pain until she gives up. I realize I should have said no to the gouging of his hand; while Garrett has not yet been officially diagnosed, I have learned a crucial lesson: do not assume health care providers know what the hell they're doing.

The nurse eventually hits the vein in his left arm, and somewhere amid the screams and tears and struggles the blood is drawn. I'm certain it wasn't just the needle that hurt. Garrett is also confused and angry. He has done nothing wrong, but it feels as though he is being punished. I try to explain, but how do you convey the enormity of a chronic, life-threatening disease to a three-year-old? How do you say that he will have to take insulin for the rest of his life, that he will be denied many foods, that he may pass out from low blood sugar, and that every organ in his body is now at risk?

"Garrett, I know it hurts, but you have a boo-boo inside your body, and we have to make it better . . . I'm so sorry, buddy, but we have to make it better."

I ask a nurse about Garrett's blood sugar. "It was high," she says, "but we don't have an exact reading." Great. Hospital's got the same damn meter I have.

I call Sheryl on my cell phone to confirm the diagnosis. Of course she's awake. "He's doing fine," I tell her. She doesn't need to know how much her son is hurting.

Garrett's night has just begun. His arm is wrapped with gauze and taped to a plank of wood, keeping the limb straight to allow an intravenous tube to be inserted. The tube is attached to a bag of saline, which will help rehydrate him until — as the doctor later explains — Garrett becomes "metabolically stable and able to eat on his own." The precaution is understandable. Most newly diagnosed children have been sick for some time and require aggressive intervention. Garrett was never so ill that he couldn't eat or drink, and he may have been fine even without the IV. But now he can't bend his arm, and he struggles to free it from the plank. The nurses also places a tube in Garrett's nose, to determine if he is exhaling carbon dioxide, which occurs during ketoacidosis and would signal the severity of his condition. The device is one more uncomfortable entanglement.

A nurse returns with information: Garrett is not ketonic, but his blood sugar is 550. I know how achy and uncomfortable I feel when my blood sugar spikes to 300. Even though he was almost twice that, he did not ask to stay home from preschool or curtail any activities. I'm sure he's been running high for weeks, but with the exception of his thirst, he rarely complained. Tough kid.

By 4 A.M., exhaustion has worn him down. He still fusses with the tube in his arm and the intolerable stick of wood, but he finally falls asleep. At some point, he is given his first injection of insulin.

Garrett had experienced pain before. Shots from the doctor. Scraped knees. A bumped head. Of course he has cried. But never this. His short life has mostly been kisses and hugs and cuddles, always very physical and affectionate, and his outpouring of love and energy was always reciprocated by those who adored him. Now he's been robbed of his childhood, I think, and he will never know what life is like without diabetes. Then again, I was diagnosed at fifteen, and I can barely remember myself.

Friends tell me later that Garrett is fortunate to have a diabetic for a father, but I'm not sure. Most parents with a newly diagnosed child fear the unknown. I knew too much.

When the sun comes up, we are in a hospital room, and Garrett is finally relieved of the tube in his nose and the IV in his arm. He enjoys pushing the buttons on the retractable bed, and television cartoons bring some relief. A nurse comes in, and I tell her that we have to be out of here in a couple of days because Garrett has a soccer game on Saturday. She pokes his finger for a morning blood sugar. "It's 279," she reports.

A weird sense of elation suddenly comes over me. This much I understand: before he got to the hospital, before he received his insulin and his IV tube and his saline, my son was dying. It wasn't imminent, but he was dying, his body unable to fulfill its most essential function — converting food into energy. Eighty-two years ago, before the discovery of insulin, he would have suffered a swift, miserable death, perhaps fading slowly at first but then rapidly and inexorably. If he were lucky, he would have been put on a starvation diet, which would prolong the agony but not change the outcome. That would have been his fate, his destiny, as it had been for countless others in the three millennia since diabetes was first recognized.

But the insulin saves him, so for the moment I do not despair over his burden but feel a surge of happiness. My son is no longer thirsty.

CHAPTER 1

Diabetic Utopia

ORLANDO, FLORIDA, JUNE 2004

At the sixty-fourth annual Scientific Sessions of the American Diabetes Association, inside the massive exhibition hall at the Orange County Convention Center, the carnival has begun.

On one stage, beneath a bright poster of grapes, bananas, and tomatoes, is "Chef Kathleen," a cheerful weight-loss guru who's busily preparing her kiwi-plum salsa for a hungry audience. Wearing a white smock, Kathleen Daelemans skillfully carves up the fresh fruit while the mahi-mahi gently sizzles on the grill. Most diabetics* feel restricted in their eating, but Chef Kathleen says that healthy diets can be delicious as long as you make the right choices. You want a taco salad? Forget about ground beef and cheese. Instead, sauté white skinless chicken breast in a nonstick pan and pile on the carrots, tomatoes, and lettuce. "Upping the good calories," she explains. Fresh seasonings are important — "If there is dust on your spice jars, they're too old" — and even flank steaks are permissible. "The key to a meat supper is portion control," she notes. While Chef Kathleen doesn't have diabetes, she says she lost seventy-five pounds, and she believes that such life-changing experiences — assisted by Web sites,

---

* Using "diabetic" as a noun is bitterly contested in the diabetes community. Many people reject this use because they believe it wrongly, even cruelly, defines a person by his or her disease. But others believe that the bias against "diabetic" reinforces the stigma, contending that the word is no different than "lawyer" or "doctor." Personally, I have never used "diabetic" as a noun, but for simplicity I will in this book, thereby avoiding the extra words in such phrases as "diabetic patients" or "people with diabetes."

chat rooms, and books, including her own — are within everyone's grasp. And she proves her point when her three slender assistants pass out samples of the mahi-mahi, the salsa spiced with a hint of Vietnamese paste.

The hall is packed with 182 exhibits, some of them adorned with twirling neon signs, gardens, or waterfalls, others equipped with interactive computer demonstrations and elaborate stages for product displays and corporate giveaways — everything from paperweights and visors to an eyeglass case that can be engraved with your name. The frills are designed to entice the convention's 13,000 attendees, most of whom care for or about diabetics — physicians, nurses, dieticians, researchers. Their recommendations can make or break a given product, and with diabetes care now costing more than $132 billion a year, glitz takes priority.

Never mind that diabetes is a complex metabolic balancing act of diet, exercise, and insulin or other medication. In these crowded aisles it can be reduced, literally, to a game. To promote its Ascensia Contour blood glucose monitor, Bayer HealthCare sets up laptop computers that display the meter on each screen. Participants navigate the device through an obstacle course called the Ascensia Challenge. The game show host stands on a circular red platform, with two "Ascensia girls" cheering from behind. To reach the finish line, the players must move their meters quickly while also destroying their enemies. Who are they? "Poor nutrition, lack of exercise, and depression are the demons of diabetics!" the host exclaims. "So if you see French fries, a couch potato, or Sad Mr. Z — zap him!" When the game begins, the Ascensia girls yell, "Blast the bad guys! Blast the bad guys!" The players take aim and fire away at all the diabetic demons. Sad Mr. Z doesn't stand a chance.

When the game ends, an impossibly handsome pitchman named Lance Porter takes a nearby stage. He has sweeping black hair with a touch of gray and a well-groomed mustache. His dark suit fits well, his tan is perfect, his voice resonant. A small crowd gathers around him. "People ask if I have diabetes," he says. "And I say, 'No, I don't' — knock on wood. But my father does." That qualifies him to speak on behalf of a company that makes glucose test strips. Most diabetics understand that controlling their disease is a lifetime struggle, but

Lance Porter believes otherwise. His book is called *28 Days to Diabetes Control!* And after those twenty-eight days, you can read the magazine that he edits: *Diabetes Positive!*

The exuberance never ends, particularly with all the goodies being served. Dippin' Dots gives samples of its "ice cream of the future." Other vendors distribute sugar-free chocolate bars, low-sugar cookies, soy-based pineapple ice cream, weight-loss milk shakes, sugarless brown sugar — all the desserts and snacks once considered taboo for diabetics but now available through the miracle of nutritional science. Of course, many of these products are still high in carbohydrates, which raise a diabetic's blood sugar; their chemicals promote indigestion, and some of the goods, such as sugar-free chocolates, taste lousy. But at least diabetics now have options. At times, however, the vendors seem to forget that diabetes is about controlling glucose levels. Takeda Pharmaceutical proclaims that its diabetes pill, Actos, "reduces insulin resistance" while it gives away chocolate crepes with bananas, whipped cream, and nuts — a dish that will increase the need for insulin.

If the food is good for diabetics, the sex is even better. In a poster for an infusion set — a small body patch and tube that attaches to an insulin pump — a topless brunette lies on her stomach, her manicured nails barely covering her breast, her blue jeans gripping her crotch, her face seductively tilted. "Want to know what comes between me and my pump?" she asks. The answer is her infusion set. "You can wear a pump and be sexy too." Certain carnal pleasures need not raise your blood sugar, either: one vendor displays sugarless sexual lubricants in kiwi-strawberry, fresh mango, and seedless watermelon.

Elsewhere, hope springs eternal for the estimated 58 percent of diabetic men who suffer from erectile dysfunction. Three different booths promote penile constriction aids — one posts a sign: "Drop your card and get a free sample" — while large signs tout the erectile dysfunction pills Cialis, Levitra, and Viagra. The maker of Cialis, Eli Lilly, was the first company to make commercial insulin in the 1920s, saving many lives and making it the most hallowed company in diabetes history. But on the convention floor, its bright green and yellow Cialis sign is far more conspicuous than anything for insulin.

Vendors proudly display therapies and treatments for various ailments that disproportionately affect diabetics. Special lubricants moisten dry eyes, cracked cuticles, and parched mouths. Unusually soft blankets and socks will not abrade the skin. Shoes with seamless soles will deter blisters. Moist heat therapy will relax muscles. A massage chair will improve circulation, and sugar-free cough drops will soothe colds. A pedometer will determine exercise levels, and Bibles on tape will provide spiritual solace to diabetics with impaired vision. Improved technology is also ubiquitous; digital imaging captures the tiny blood vessels of the retina while software programs help patients track their glucose levels. The meters are now so fast and efficient that their manufacturers emphasize their ergonomically designed contours and waterproof casing.

Novo Nordisk decides to take a page out of *Architectural Digest* for its exhibit. It has constructed a modular two-level condo spread across 10,000 square feet, with inlaid floor lights, huge video screens, small trees in white brick planters, glassed-in meeting rooms, sleek computers, and a cobalt-blue fish tank. Attractive blond "facilitators" greet visitors while a rugged mountain climber and diabetic named Will Cross chats about his latest trip up Mount Everest. He smiles for photographs. Company officials are on hand to discuss their insulin, but the rush of traffic occurs at lunch, when they serve blinis with caviar, salmon, and sour cream. A similar frenzy erupts over an afternoon snack of whipped vanilla ice cream with cherries and almonds.

The word in the hall is that Novo Nordisk spent about $1.8 million for the entire exhibit — what one rival, suffering from admitted booth envy, says "would make a nice addition to my house" — or, for that matter, a new house entirely. A company official won't comment on the cost but says the exhibit shows that Novo Nordisk is "an innovative frontier company."

This fantasyland puts nearby Walt Disney World to shame, a diabetic utopia that is also reproduced in heartwarming television commercials, sunny magazine ads, and glossy marketing brochures — a biomedical golden age in which genetics, immunology, and stem cell biology are unlocking the secrets of this disease.

These notions have some basis in truth. Our understanding of diabetes has indeed improved, yielding better treatments, improved technologies for complications, more aggressive interventions, less painful therapies (including inhaled insulin), and smarter guidance for its general management. Many diabetics enjoy long, productive lives and believe, rightly, that their disorder does not limit them in any way. Some even recoil at the word "disease" and insist they have "a condition." Anyone who develops diabetes today has access to therapies unimaginable a generation ago; and the *opportunity* of living your natural lifespan (as opposed to *likelihood*) is better than at any time in history.

But that very progress highlights the crushing contradictions of the disease. While its treatment has become more sophisticated, overall care is not improving, and by many measures it is getting worse. While the most common form of diabetes can often be prevented through weight loss, nutrition, and exercise, the disease itself has become America's biggest epidemic. While the risk of complications to the eyes, heart, nerves, and kidneys can be greatly minimized, their incidence is only growing. While the federal government and fundraising organizations launch publicity campaigns about diabetes, misperceptions still run high. And while scientists drill down to its molecular underpinnings, its cure remains maddeningly elusive.

Thirty years ago, diabetes was a relatively rare disorder that mainly affected the elderly; in more severe cases involving children, it could be controlled with insulin. If anything, diabetes was a success story in which the elixir, insulin, restored life to the doomed, a modern parable affirming scientists' ability to master a deadly illness. The title of a 1962 book — *The Story of Insulin: Forty Years of Success Against Diabetes* — makes the point. The author writes: "Few stories of discovery carry more drama than that about Insulin. The leap from despair to death was so sudden . . . More and more people of all ages have had the moving experience of being drawn back from sickness and death into health and happiness."

This narrative was often played out in news stories about diabetic children at summer camp. The first such camps were founded shortly after insulin was discovered, in 1922, leading to dozens more around the country, and newspaper and television reporters could count on them for an uplifting summer yarn. It would feature happy, robust children swimming in the lake, playing soccer, and singing camp songs, as well as injecting their insulin, weighing their food, and, in more recent years, testing their blood sugar or adjusting their insulin pumps. As a counselor myself in the late 1970s and early '80s, I participated in several of these stories, and they reinforced the notion of diabetes as an isolated and manageable disorder, an inconvenience to its doughty victims but a peripheral concern to the public.

Today the story is far different. Diabetes is America's most common and costly chronic illness, touching all segments of society while ravaging minority communities in particular; in some cases, it afflicts three generations of a single family. In the largest U.S. city, New York, an estimated 800,000 people, or more than one in every eight, have diabetes. Across the nation, it's the only major disease whose death rate is rising, up 22 percent since 1980. The crisis has also spread abroad. India and China are now the global leaders in diabetes, and developing countries in general, which rarely saw the disease thirty years ago, now face an epidemic: a rise in sedentary lifestyles and obesity, fueled by industrialization and Western diets, has made them vulnerable to the most common form of the disease. Few countries have been spared. According to the World Health Organization, the number of people worldwide with diabetes has increased from an estimated 30 million in 1985 to 194 million in 2003 — to a projected 333 million by 2025. In two generations, the disease will increase by more than 1,000 percent. Paul Zimmet, director of the International Diabetes Institute in Melbourne, Australia, says the disease "is one of the main threats to human health in the twenty-first century."

Diabetes has often been called an "invisible" disease because its burdens are not expressed outwardly and because its patients have sought to hide it. But it's invisible no longer. It's featured in TV commercials and consumer magazines and is incorporated into movies, sitcoms, and novels. Supermarkets and drugstores have entire aisles devoted to appropriate foods and health care items. There is even a weekly cable show devoted to the subject, accompanied by a Web site, podcasts, and blogs. Diabetes has attracted famous spokespersons, such as the actresses Halle Berry and Mary Tyler Moore, the jazz legend B. B. King, and the former Miss America Nicole Johnson Baker; it has also claimed many well-known casualties, such as Jackie Robinson, Thomas Edison, and Johnny Cash.

For all the increased attention, misperceptions abound, in part because diabetes comes in several different but related forms. In every case, the body is unable to convert food into energy, leading to chronic high blood sugar if left untreated. In the nondiabetic, food is digested and carbohydrates are broken down into glucose, which provides energy for cells. The glucose travels to the liver, which is programmed to save some of the substance while releasing the rest into the bloodstream. When blood glucose levels begin to rise, the pancreas senses the sugary tide and releases insulin, which serves as a key, opening the cell doors and allowing the glucose to enter. The more pasta or cereal or juice consumed, the more insulin released, producing quick spikes in the blood sugar, which soon returns to a normal range. The whole process is a remarkable metabolic symphony, fluid and seamless, each movement leading inexorably to the next until glycemic order is restored.

The process breaks down in diabetics. Most cases fall into one of two overlapping categories. The kind that affects about 90 percent of patients, type 2 diabetes, arises when the body does not respond to the insulin secreted by the pancreatic beta cells: faulty "insulin receptors" prevent the key from opening the cell door. "Insulin resistance" is quite common among those who are obese, sedentary, or aging. Most of them, however, do not become diabetic — they can produce additional insulin to maintain normal blood sugar. Type 2 diabetes only occurs when the beta cells are also impaired. Once known as "adult-onset diabetes," its name changed because increasing numbers of heavy teenagers developed it. Treatment typically consists of diet, exercise, and oral medication, though about a quarter of these patients take insulin. Experts say that twice as many should be taking insulin but aren't, either because they fear the injections or because their doctors don't understand the disease.

The second most common form of diabetes, type 1, occurs when the pancreas shuts down all or most of its insulin production, forcing the patient to replace it through injections or some other delivery system. Formerly called "juvenile onset," it is an autoimmune disorder, meaning that the body's own immune system ambushes the beta cells. What makes this disease so infuriating is that the immune system exists for only one reason: to protect the body, to shield it from foreign substances, viruses, and other harmful organisms. But in diabetics, the immune system attacks the very cells that it is supposed to defend — like palace guards storming their own garrison or Secret Service agents turning on their president. The diabetic body endures an act of rank perfidy.

While these two types of diabetes have clear medical differences, they are both characterized by the loss of beta cells. Type 2 patients experience the loss more gradually but inexorably, and they will eventually need insulin to survive, just like their type 1 counterparts. The burdens are also similar, requiring a daily juggle of diet, exercise, and medication to maintain a normal blood sugar — which is to say, an unusual amount of self-discipline and personal initiative. Diabetes in any form subverts the doctor's traditional role, ultimately placing the patient in charge. Both types are influenced by genetics, although type 2 more than type 1. Both are costly to treat and unforgiving if ignored, exacting the same revenge in vascular complications. And both carry an emotional toll. Many patients are frightened by the disorder's long-term effects; they feel anger and shame about having the disease and go to considerable lengths to conceal it.

At least two other types of diabetes exist. One is gestational: it affects about 4 percent of pregnant women and appears to be caused by the hormones of the placenta, which block the action of the mother's insulin in her body. Another form, identified only in the past few years, has features of both type 1 and type 2: the body doesn't make its own insulin but resists injected insulin as well. Called "double diabetes" or type 1.5, it is "the consequence of [having] type 1 diabetes in a progressively obese population," said Philip Zeitler, an associate professor of pediatrics at the University of Colorado Health Sciences Center in Denver.

For all the media attention diabetes has received in recent years, the true scope of the problem has been understated.

Calculating the number of diabetics in America has always been a problem because there is no national registry or database, which is more typically used with infectious diseases. But estimates confirm a significant increase. The Centers for Disease Control and Prevention (CDC), relying on various surveys, estimates that in 1990, 4.9 percent of adults in America had diabetes; by 2002, that number had almost doubled to 8.7 percent. Between 1996 and 2005, the number of actual diagnosed cases had increased by 71.8 percent, from 8.5 million to 14.6 million. The CDC believes that, including undiagnosed cases, 20.8 million Americans have diabetes, a figure widely reported in news accounts. But according to the CDC, 20.8 million is the estimated *minimum* number. The real figure is greater. "We have a passion not to overestimate numbers because it could lead to inappropriate findings," said Jane Kelly, the program director for the CDC's National Diabetes Education Program. "We can say that number is the minimum that we estimate. Who knows what the maximum might be? Then you would enter into the realm of conjecture."

Such accounting rankles diabetics and their advocates, who believe a true measure of the disease would justify greater funding for research. The number of children with diabetes is even murkier. Until several years ago, essentially no effort was made to count diabetic youths. Then, in 2000, the CDC announced a million-dollar study to count representative samples of those children; preliminary estimates indicate that about 150,000 youths under age eighteen have type 1 diabetes, or one in every 400 to 500. The total number, of course, would be higher if type 2 children were included.

The death rate from diabetes is also understated. Citing the CDC, news reports often describe the illness as the sixth leading cause of death in America, but they misrepresent the CDC's findings. The agency's ranking was based on a survey of death certificates in 2000, which showed that diabetes was the underlying cause of death in 69,301 cases. But the disease actually contributed to 213,062 deaths. Moreover, the CDC concluded that "diabetes was likely underreported as a cause of death," for studies have shown that only about 35 percent of decedents with the disease have it cited anywhere on their death certificate, and only about 10 percent to 15 percent have it noted as the underlying cause. Not all diabetics, of course, die from diabetes, but the underreporting downplays its true impact.

High death rates should not be surprising, given the overall level of diabetic care. According to the National Institute of Diabetes & Digestive & Kidney Diseases, which is part of the National Institutes of Health (NIH), fewer than 12 percent of the diabetics surveyed met the recommended goals for blood glucose, blood pressure, and cholesterol, and overall diabetic control has deteriorated. Between 1999 and 2000, 37 percent of those surveyed achieved average three-month blood sugars below the American Diabetes Association's recommended level compared to 44 percent in the same survey between 1988 and 1994. Even patients at leading research hospitals are falling short. In one study involving thirty academic medical centers, only 10 percent of diabetic patients met the recommended goals for blood sugar, blood pressure, and cholesterol levels, and physicians often failed to provide proper guidance to poorly controlled patients. According to the study, fewer than half of those patients were instructed to adjust their regimen.

In 1897, the pioneering British physician Sir William Osler said, "Know syphilis in all its manifestations and relations, and all other things clinical will be added unto you." That could now be said of diabetes. The disease is effectively present wherever blood flows, penetrating all corners of the body, preying on vascular cells, sandbagging organs, dulling nerves. So toxic is the glucose that when one child was diagnosed, the nurse told his mother that the lab results indicated his blood "was incompatible with human life."

Diabetes increases the risk of heart disease by six times and strokes by four, and it makes Alzheimer's, asthma, and hypertension more likely. Each year, the disease blinds 24,000 Americans, making it the leading cause of lost vision among adults, and it cripples the kidneys of 28,000 patients. It is also driving the increase in amputations. In 2002, more than 110,000 lower extremities were amputated, up from 99,522 in 1993, according to the Agency for Healthcare Research and Quality. Diabetics account for more than half of all lower-limb amputations.

Research also suggests that hyperglycemia — high blood sugar — is more damaging than once thought. Scientists used to believe that atherosclerosis, or the plaque inside arterial walls, rarely occurred in type 1 diabetes. But in a study of diabetics, not a single participant had normal blood vessels. The average age of the group was only forty-two, but all twenty-nine subjects, even those with the best glycemic control, had atherosclerotic plaque, which increases the risk of heart disease. It appears that for most diabetics, it's not whether this complication will occur but when and how severely.

Scientists can quantify the number of years that diabetes will cost an average patient as well as measure the quality of those years. According to the CDC, men diagnosed by the age of forty will lose more than eleven years; women will be denied fourteen years. When calculated for "quality-adjusted life years," the men give up nineteen years of quality life; the women, twenty-two.

"Clinical inertia," or the failure to provide intensive therapy, leads to many unnecessary outcomes. More than 14,000 heart attacks, strokes, and amputations could be prevented each year through better diabetes management, according to the National Committee for Quality Assurance, a nonprofit organization that measures the quality of health care. Diabetes has "higher avoidable hospital costs" than any other disease — $573 million a year — and substandard care leads to as many as 9,600 "avoidable deaths" a year. That is higher than the figures for all but two other diseases, high cholesterol and high blood pressure.

The cumulative toll of the disease amounts to little more than a skein of bloodless numbers; its true impact can be found in the experiences of someone like Karen Katsamore. Diagnosed in the late 1940s at the age of four, she grew up in Queens, New York. She wanted to be a ballroom dancer, but she had to manage her illness in the dark ages of the insulin era, without the means or the knowledge to test or control blood sugar. Now in her sixties, she says she has "gone through most of the horrors this silent killer can impose on the human body": both legs amputated, a kidney transplant, two heart attacks, a triple heart bypass surgery, and four cataract surgeries. But none of those sorrows compares to her heartbreak in trying to bear a child. When her first pregnancy miscarried after four months, her obstetrician recommended tubal ligation, but she refused. Her second pregnancy progressed better. The first two trimesters passed without incident. She often felt tired, though she attributed that to her growing baby as much as to her diabetes. Her doctor, however, remained oddly sullen and low key. At seven months, her daughter was stillborn. In the days that followed, the doctor said that only a miracle could have delivered that baby to term, and he looked relieved because he had feared that Katsamore herself would die. He said she might not make it the next time, so she agreed to have her tubes tied. Many years later she wrote, "My only regret in all this is that I never asked to hold my baby when she was delivered. I didn't know at the time I could do that and it still haunts me to think I let that opportunity pass me by. To carry life inside you and feel it moving for months and to want to take a glimpse of someone two people in love have created, is the most beautiful thing in the world. I would have carried that memory to my death but I let it pass me by unknowingly."

The financial toll of diabetes is also mounting, though it too is understated. The media often report that the medical cost for diabetes is $132 billion, but that figure, estimated by the American Diabetes Association (ADA), applied to 2002. Given the spiraling growth rate — experts estimate that more than 4,000 new diagnoses are made a day — the annual costs are certainly much higher. While the ADA's figure includes some indirect costs, such as lost workdays, it ignores many others. The time burdens of the disease — testing blood sugars, taking medication, waiting in doctors' offices — is an obvious example; less obvious is gasoline. Debra Hull has spent most of her life in isolated farming communities. Now, at thirty-seven, she lives in northwest Missouri; she estimates she has spent $8,500 in gas going to pharmacies, doctors' offices, and hospital labs.

The high cost of diabetes also reflects the improved treatments. Expensive tests, medications, and therapies, including insulin pumps, are receiving greater acceptance, and more sophisticated technologies, such as continuous glucose monitors, are reaching the market. Even using the ADA's dated, conservative figures, the annual health care cost of a diabetic is more than five times that of a nondiabetic — $13,243 compared to $2,560. About 18 percent of Medicare beneficiaries have diabetes, but they account for 32 percent of Medicare spending. Test strips alone represented almost 10 percent — about

$740 million — of Medicare's total health care expenditures in 2002, according to the Centers for Medicare and Medicaid Services.

Ironically, the high price of care complicates the ability of diabetics to get the insurance they need to stay healthy. A survey of 851 patients indicated that many health insurance policies do not cover basic diabetic needs, and preexisting exclusions often deterred them from receiving coverage, forcing them to join expensive, high-risk pools. The survey, sponsored by the ADA with the Georgetown University Health Policy Institute, also showed that diabetes is still deemed "uninsurable" by some medical underwriting practices.

Behind the epidemic are powerful social forces — obesity and the aging of the population — that will continue to drive the crisis. In 2004, the government estimated that 41 million Americans have "pre-diabetes," blood sugar high enough to put them at substantial risk. The CDC believes that one in three people born in 2000 will develop the disorder. What's more, many people are developing it at younger ages, giving them a much higher probability for complications. (Someone who contracts diabetes at the age of seventy typically doesn't have time to develop them.) This trend has frightening consequences for the health care system. About 75 percent of the direct medical costs for diabetes are associated with treatment for late microvascular, neuropathic, and cardiovascular complications. Tommy Thompson, the secretary of health and human services during President George W. Bush's first term, warned that the country "risks being overwhelmed by the health and human consequences of an ever-growing diabetes epidemic."

That diabetes is our most daunting public health threat underscores deeper problems in America's $1.8 trillion health care system. Sprawling, fragmented, and inefficient, the system was designed to treat acute or episodic illnesses, rewarding physicians who could swiftly intervene to identify a disease or injury, arrest its progress, and repair the damage. This system, in most cases, generously compensates specialists, such as surgeons, cardiologists, and radiologists, the more so if they use sophisticated equipment for invasive procedures. But it poorly serves a country in which chronic disease is far more costly and prevalent than acute illness. Relatively little is spent on keeping people healthy through preventative care while enormous sums are paid when they become sick or disabled.

Nowhere is this misalignment more evident than in diabetes, which, given its physical and emotional complexity, stands as the quintessential chronic disorder. On one level, a physician must understand how to apply a war chest of drugs and therapies, often in tandem with those for other illnesses, and what risk factors to examine and identify. But treatment also requires discovering the underlying issues — social, emotional, familial — that will ultimately determine a patient's medical outcome. This requires cognitive skills, introspection, and time with the patient, but these tasks, compared to others in medicine, are not well compensated. Most physicians or clinics who try to keep diabetics healthy lose money.

Which is not to say that diabetics are bad for the medical business. Once complications occur, patients need hospitalization, amputation, coronary bypass surgery, dialysis, laser photocoagulation, or referrals to the many specialists who care for vascular problems. In this sense, only after the diabetic body becomes damaged does it become economically viable to treat, reflecting a reimbursement system that actually rewards poor care. "The surgeon who lops off a leg of a diabetic patient gets paid a lot, but the endocrinologist who keeps the leg healthy is paid relatively little," said Kenneth Ludmerer, a medical historian at Washington University.

Diabetes also dramatizes the growing divide between the haves and the have-nots in medical care. Those with financial resources can use increasingly expensive therapies — from $6,000 insulin pumps to new injectable agents like exenatide (Byetta), which costs about $8 per day. These, of course, are beyond the reach of the uninsured and the unemployed in America's outcast system of public health. Even those with coverage discover that their insurers will not always pay for the best treatments. What is emerging is an elite corps of diabetics — highly motivated, educated, and financially secure — who are flourishing in the very world imagined at the ADA conference in Orlando, compared to the nearly 90 percent who fail to meet basic goals for blood glucose, blood pressure, and cholesterol levels.

For all the hype and hucksterism, the tragedy of diabetes is that no other chronic disease has the "exquisite epidemiological indicators" in demonstrating how to reduce the risks for complications and morbidity, according to Jeremy Nobel, an adjunct lecturer at the Harvard School of Public Health. Clinical studies have demonstrated how to reduce bad outcomes, and the necessary tools are available. If the health care providers, the payers, and the patients themselves were all committed, then diabetes would be a clinical success story.

It is not the worst disease in the world, far from it, but diabetes carries one other attribute that has always given it a tragic and poignant subtext: it affects children. We accept disease as part of aging, but there is no accepting a disorder that afflicts the most vulnerable and innocent among us, no justice in a world that would rob a child of childhood. This has led to extraordinary efforts to find a cure, but those efforts have put patients and their loved ones on an endless emotional roller coaster. Every disease, in the search for a cure, generates a sine wave of great expectation followed by disillusionment; but in diabetes, the hopes have been higher, the promises more explicit, the disappointments more bitter. Unrealistic expectations have been fueled in part by self-promoting scientists and credulous reporters, but the distinctive history of this disease has also played a role. If the miracle of insulin was seen as a life-saving partial cure, how difficult could it be for scientists to simply finish the job?

Very, it turns out. Scientists have failed for many reasons — from the inscrutability of the immune system to the shortcomings of our medical research industry — but their failure hasn't extinguished the dreams of those who have the most to gain. As Gail O'Keefe, whose two children both have diabetes, wrote to me: "We wish for a cure, yes, because it is a difficult road. Because it is impossibly sad to draw blood eight times a day from a seven-month-old infant. Because it is enormously frustrating to do all you can, only to have technology fail you and an infusion set fall out at the soccer tournament. Because to imagine your beautiful thirteen-year-old with a shortened lifespan, amputations, or blindness weighs endlessly on a mother's heart. Because when you visualize the tiny drop of insulin that represents what is missing in your loved one's body, it seems too trivial. Scientists can do so much — how is it they cannot coax the cells to make this tiny drop of protein on demand?"

And sometimes in diabetes, a combination of faith and denial endures long after the cause is lost — as it does for Marilyn Cattanach, whose son, John, died from diabetic complications in 2000. In the refrigerator in her house in Wellesley, Massachusetts, she keeps his last insulin bottle. "I don't think," she says, "that they really understand this disease."

CHAPTER 2

Insulin's Poster Girl

FROM HER EARLIEST DAYS, Elizabeth Evans Hughes had a front-row seat to history. Born in 1907, she was the daughter of Charles Evans Hughes, the governor of New York and a rising star in the Republican Party. The family entertained America's First Couple, allowing President William Howard Taft to bounce Elizabeth on his ample knee. In 1910, the president appointed Hughes to the U.S. Supreme Court, a position he held until 1916, when he became a candidate for president. The campaign excited Elizabeth, who was already familiar with her possible new living quarters. "If father is elected and we live in the White House, I can bathe in the fountains," she said. She almost got her wish, but her father lost to the incumbent, Woodrow Wilson, in one of the closest elections in American history. When the Republicans regained the White House in 1920, Hughes served as secretary of state for eight years under two presidents. He returned to the Supreme Court in 1930, this time as the chief justice, until he retired in 1941. His remarkable career allowed Elizabeth to meet luminaries in politics, law, and business. When the British statesman Lloyd George came to visit, Elizabeth chatted with him for five minutes. "She walked out on air, as only a girl of sixteen can," wrote Charles Hughes's biographer.

But Elizabeth had her own moment of fame, attracting international attention and, ultimately, writing one of the more poignant chapters in medical history. When she was eleven, her body began to fail. Inexplicably hungry and constantly thirsty, she would come home from birthday parties where she had eaten ice cream and cake but now craved water, sometimes drinking two quarts. Never a big girl, she began to lose weight even as she devoured larger quantities of food. Her upbeat spirit gave way to exhaustion. Her parents scrambled for help.

In the spring of 1919, Elizabeth was diagnosed with diabetes. With access to the finest doctors in America, she was sent to Morristown, New Jersey, where a physician named Frederick Allen had won acclaim for an unusual therapy that could prolong the lives of diabetics. He told Elizabeth's parents that she could live for another year, three at the most. That seemed like a small blessing, but the actual treatment was hardly reassuring. In fact, Allen's cheerless personality befit the care. Though Elizabeth weighed only seventy-five pounds and was losing weight, Allen did what he did best. As one journalist wrote, "He began to starve her."

Frederick Allen's therapy, however perverse, still represented a thin reed of hope after almost three thousand years of frustration and despair with the disease doctors called the "pissing evil." The Greek physician Aretaeus of Cappadocia offered the first accurate account of diabetes in the first century A.D., noting the distinctively gruesome fashion in which some patients withered away. He called their demise "the melting down of the flesh and limbs into urine." He continued, "The nature of the disease is chronic, and it takes a long period to form; but the patient is short-lived; if the constitution of the disease be completely established, for the melting is rapid, the death speedy . . . Life is disgusting and painful." In the 1600s, an English doctor made the disorder's first connection with sugar when he recognized the patient's urine was sweet, "as if imbued with honey or sugar." This observation led to the malady's complete name, diabetes mellitus. The Greek prefix "*dia*" means "to pass through," while "*betes*" is "tube," so "diabetes" means "water siphon." "*Mellitus*" is "sweetened with honey."

Sweet urine had practical benefits: a doctor's taste test could diagnose the disease, but it appears that some patients had to experience the ignominy of their own diagnosis. Elliott P. Joslin, who began treating diabetes in the 1890s, reviewed the records of every such patient who entered Massachusetts General Hospital from 1824 to 1898. He concluded that while a physician would taste a diabetic's urine, "it is probable that a patient occasionally analyzed his own," so he knew at first hand the consequences of overeating: "the urine was increased correspondingly in sweetness." Even after chemical analysis had replaced the taste test, physicians would still detect a male diabetic by observing white spots of dried sugar on his shoes. An incontinent diabetic on his deathbed would sometimes attract black ants.

Diabetes's signature symptom was polyuria, excessive urine: when blood sugar levels rise, the body draws water from its tissues to purge the sugar through its urine. In 1727, the English poet and physician Sir Richard Blackmore described the copious flow "as when the Treasures of Snow collected in Winter on the Alpine Hills, and dissolved and thawed by the first hot Days of the returning Spring, flow down in Torrents through the abrupt Channels, and overspread the Vales with a sudden Inundation." The clinical translation: in 1861, a patient at Mass. General voided 36 pints in one day.

Treatments failed, in part because scientists knew so little about disease in general. Until the middle of the nineteenth century, illnesses were viewed as an imbalance in the body, to be offset by "natural processes" such as urinating, sweating, or vomiting. A doctor's job was to accelerate these outflows, the most popular being bleeding. All were tried on diabetics, as were various diets (oats, milk, rice); opium was also used. But nothing could reverse the diabetic's decline.

The disorder affected the entire body, impairing its ability to heal from cuts or wounds. Patients were more vulnerable to tuberculosis and pneumonia. Boils, carbuncles, and gangrene often had fatal consequences, and any surgery was believed too risky. If complications didn't doom the patient, the disease ran its own pernicious course. As the body failed to metabolize its food, it called on fatty acids for energy, which led to a buildup of a chemical called ketone. Over time, ketone bodies clogged the bloodstream and passed out through the urine. Some ketones were breathed out, creating a sickish apple smell that hung like death in hospital wards. By then, food and water were irrelevant, the downward spiral irreversible. As the ketones accumulated, the body's pH level declined to dangerously acidic levels — a state known as diabetic ketoacidosis. As the blood became more

27

acidic, the concentration of bicarbonate fell and more carbonic acid was formed, which the patient tried to expel through deep, rapid breaths. As he slipped into a coma, his chest would heave in a desperate attempt to push out more of the gas, a process sometimes known as "air hunger" or "internal suffocation." Death was only a few hours away.

Scientists made little headway in understanding the disease, let alone in curing it. As a British doctor said of a French diabetologist, "What sin has Pavy committed, or his fathers before him, that he should be condemned to spend his life seeking for the cure of an incurable disease?" Diabetics were hopeless, medical pariahs. In 1914, Joslin noted that for every hundred diabetics admitted to Mass. General, twenty-eight "are discharged dead. . . . Surgeons dodge the diabetic, while the obstetrician is out and out afraid of diabetes and urges pregnant women to have abortions. The neurologist, dermatologist, and ophthalmologist will throw up their hands at complications . . . and exclaim, 'Cure the diabetes and then we will help the patient.'"

As often happens in science, a breakthrough occurred more by happenstance than design. In 1889, the German physiologists Oskar Minkowski and Joseph von Mering were trying to settle a debate on whether pancreatic enzymes were needed to digest fat. They removed the pancreas from a dog, which unexpectedly caused increased urination. Minkowski tested the urine and found sugar, leading him to identify the pancreas as the source of an "antidiabetic" substance. The pancreas, a long, lumpy gland beneath the stomach, is nestled in the hollow horseshoe curve of the duodenum, which is largely responsible for the breakdown of food in the small intestine. Scientists already knew that the pancreas had another function. Thirty years earlier, a Berlin medical student named Paul Langerhans had identified clusters of cells in the pancreas that did not secrete the normal pancreatic juice but whose purpose was unknown. These mysterious cells, or islets, would be given the lyrical name "islands of Langerhans."

By the early twentieth century, researchers had connected these islets with Minkowski's discovery, showing that damaging the islets produced diabetes. This connection prompted a frenetic search for a "pancreatic extract" that could lower a person's blood sugar. Until they could find such a substance, there was only one way to lengthen the life of a diabetic, and that was starvation.

Frederick Allen was not the first person to recognize the importance of carbohydrates in metabolism. In 1822, a U.S. army surgeon, William Beaumont, was summoned to the Canadian border to help a nineteen-year-old trapper who had a gaping shotgun wound in his abdomen. The youngster recovered, but he had a hole in his stomach that had to be plugged when he ate to keep the food from coming out. Beaumont used this unfortunate result to examine digestion, tying pieces of food to silk string and dropping them into the victim's stomach. He discovered that salted pork and beef were digested slowly, but stale bread broke down the quickest. Diabetes's connection with food was strengthened by a French doctor, Apollinaire Bouchardat, who in 1870 was experimenting with periodic fasts for his patients. When Paris was attacked by Germany, food rationing imposed even greater restrictions, and Bouchardat noticed the disappearance of sugar in the urine of diabetics. Exercise, he observed, also seemed to increase their tolerance for carbohydrates.

But Frederick Allen did not believe simply in cutting carbohydrates. His therapy required cutting protein and fat as well — anything that could produce diabetic symptoms. "All food," he wrote, "contains danger." He tested his theory on diabetic dogs and cats, publishing his results in an epic report (1,179 pages) in 1913. He began experimenting on humans the following year. While he described his therapy in terms of diet and nutrition, others were more blunt. In 1917, a book about this treatment by a physician and a dietician was titled *The Starvation Treatment of Diabetes.*

On one level, the approach probably seemed logical. Allen, like other diabetologists, had noticed that many patients were overweight — "Luxurious living and sedentary life are thought to predispose to this disorder," he wrote — so a punitive diet seemed appropriate. Patients were initially denied all food until their urine was free of sugar, which could take four days. Then they were gradually given sprigs of food, such as "thrice-boiled" vegetables, the water discarded after each boil to eliminate any remaining calories. Patients would add to their diet — an egg here, a piece of fish there — to determine their tolerance. Once sugar reappeared in their urine, another fast was imposed. Finally, the patients would be allowed a minimal diet that took them to the brink of spilling sugar into their urine. Doctors tried to dull the pain through alcohol. During the treatment's first phase, patients were given black coffee with one ounce of whisky every two hours. "It merely furnishes a few calories and keeps the patient more comfortable while he is being starved," a nurse wrote.

Doctors hailed the diet's ability to keep patients alive, however briefly and uncomfortably, and no one was more supportive than Elliott Joslin. America's preeminent diabetologist had long emphasized nutrition, taking patients to his house in Boston, where he would feed them dinner and instruct them on proper eating habits. Armed with Allen's diet, he began to quantify how many more months his patients were living. "There was a thrill in the knowledge that undernutrition had changed the prognosis and that length of life of the fatal cases had been six years instead of 4.9 years," he later recalled.

But the therapy raised a more fundamental issue, which framed diabetes care for the rest of the century. If patients could prolong their lives through diet, then they could be blamed if the diet failed. Such patient responsibility gained much more traction after insulin's discovery. (The German physician Bernard Naunyn had his own way of enforcing patient compliance: he locked patients in their rooms for five months to obtain "sugar freedom.") But Allen began a clear shift in transferring the burden to the patients themselves. The problem, of course, was that the diet was cruel. While some diabetics were overweight, many others, including all the children, were already emaciated and deteriorating, craving the one thing they were being denied: food.

Allen's clinic, the Psychiatric Institute, became a famine ward, or worse, and patients made desperate attempts to gain nourishment. When a blind twelve-year-old boy was isolated in a hospital room, he asked for a canary and secretly ate birdseed from the cage. He also consumed toothpaste. These efforts, which caused him to spill sugar in his urine, elicited scorn from Allen. He later wrote, "These facts were obtained by confession after long and plausible denials." His food restricted, the boy weighed less than forty pounds when he died of starvation. One nurse said, "It would have been unendurable if only there had not been so many others."

Both Joslin and Allen believed that death from starvation, or "inanition," was somehow more benign than a conventional diabetic death. "Inanition will undoubtedly increase in frequency, but I shall not allow that to divert my attention from the 60 per cent of cases who die of coma," Joslin said. "The one enemy which the diabetic must fight is coma." Allen said that inanition could only be avoided by starving patients earlier — "thorough undernutrition at the outset." When that failed, he was inclined to blame the patients or their families, not the disease itself. He wrote that death from "active diabetic symptoms" is "produced by lax diets or by violations of diet," citing case studies to prove his point, such as that of one sixteen-year-old boy whose decline was precipitated by a bowl of cherries.

Elizabeth Evans Hughes's diagnosis, in April of 1919, coincided with another family tragedy. That very month her twenty-eight-year-old sister, Helen, contracted tuberculosis. One can only imagine the grief of parents who saw two of their four children develop a fatal disease in the same month. Helen died the following year. Having buried their oldest daughter, Charles and Antoinette Hughes now watched their youngest cling to life.

Elizabeth herself never knew how serious her condition was, only the imperative of maintaining her diet. When she was admitted to Allen's clinic, she was four feet, eleven inches, and weighed seventy-five pounds. Allen starved her for about a week, then gradually increased her diet, reaching 70 grams (2.5 ounces) of carbohydrate a day, including thrice-cooked spinach and cabbage. She still fasted one day a week, though she might be allowed an egg in the morning and some chicken at lunch. Her weight dropped to fifty-five pounds. Not surprisingly, she developed an intense dislike for both Allen — whose jowly face and cold demeanor reminded her of a bulldog — and his uncompromising diet. She ate lean meat, lettuce, milk, bran rusks, and eggs, sometimes scrambled, sometimes raw. The regimen did indeed make her "sugar free," and she impressed Allen, who called her "a model patient." She appears to have stayed in the clinic for only a few weeks but would return periodically. Without Allen's care, her parents hired a nurse to prepare her meals and vigorously enforce the rules. On Thanksgiving, Elizabeth snuck into her room with a little piece of turkey skin, but she was roundly chastised. "You must never take extras," the nurse warned.

Elizabeth's parents hoped that better weather would improve her health, so she and her nurse were sent on trips to the Adirondacks, in upstate New York, and even to Bermuda. Her letters home convey the optimism of a teenage romantic, impervious to her fatal condition while gushing over the nature writings of John Burroughs, the splendor of a bald eagle, or the enjoyment of night fishing. "What is more fun than camping anyway?" she wrote. "Close to nature with the sky for the roof and Mother Nature for a floor." All she wanted for Christmas was a canary.

When sugar appeared in her urine, her food was reduced; but she did not complain, even when she was eating the equivalent of less than one piece of bread a day. She learned to adjust, to endure, to simply make do. She bought a white dress with an extra piece of lace on the collar to make the garment look broad and soft. "The other way made me look too much like myself: narrow and skimpy," she wrote. She took pride in exercising each day, no matter how tired she was. She was sensitive to her otherness, describing how her camping friends indulged in sandwiches, salads, and watermelon while she ate roasted chops and baked custard. But she embraced accommodations that could have been painful or embarrassing as her unique identity. On her fourteenth birthday, she received a hatbox covered with pink paper and fifteen candles, which she joyfully blew out.

While Elizabeth's starvation diet had indeed prolonged her life, so too had her own buoyant spirit and fierce determination. Her letters no doubt glossed over some of her pain to protect her parents, but they also exude her joy for life and her refusal to even acknowledge the seriousness of her ailment. Nonetheless, by the summer of 1922 she had reached the outer limit of her life expectancy with diabetes — three years — and her end was coming near. Illnesses, such as colds and tonsillitis, had set her back, forcing constant adjustments in her food. In the spring, she had an ulcerated tooth and diarrhea; while in Bermuda, she was confined to her hammock, where she would read, write, or knit. She was eating fewer than three hundred calories a day. "I hope we can avoid any more such [illnesses] because she has so little reserve strength," her nurse wrote. When Elizabeth boarded the ship to return home, she insisted on walking up the long ramp, despite her weakness. She weighed less than fifty pounds, and it appeared that her parents could do little more than wait to bury their second child in three years.

Then Antoinette Hughes saw an item in the newspaper about an experimental therapy for diabetes being developed in Toronto. She called a physician at Johns Hopkins University who said, "By all means take her up there. It can't hurt." On July 3, 1922, she wrote to Frederick Banting at the University of Toronto, informing him that her daughter was in a "weak and wasting condition." She was not necessarily seeking a cure for her child. She just wanted a treatment that would allow Elizabeth to eat more food — to "increase her tolerance" — so her remaining days would not pass on an empty stomach.

Fred Banting could not have been a more unlikely figure to forge one of the great breakthroughs of medical history. Born on a small farm in Ontario in 1891, he was a serious but average student who drifted into medical school at the University of Toronto. Serving as a medical officer in World War I, he was wounded in France and received a Military Cross from the British government. He then returned to Ontario, where he struggled to make a living as a country surgeon. He accepted a part-time teaching position at a medical school and in October 1920 was asked to give a lecture on a topic he knew little about: carbohydrate metabolism and its associated disease, diabetes.

To prepare, he read an article in *Surgery, Gynecology and Obstetrics* by an American pathologist who had come across an unusual finding during a routine autopsy. He noticed that a pancreatic stone had obstructed the main pancreatic duct, and that while most of the enzyme-secreting acinar cells had disappeared through degeneration, the islet cells had apparently survived. His review of the literature found animal experiments in which the ducts were bound, leaving behind healthy islets but no sugar in the urine, and his finding seemed to reinforce the idea that the islets played an essential role in diabetes.

This notion spurred Banting to consider a possible diabetic treatment: tying a dog's pancreatic ducts to isolate the organ's "internal secretion," the extract that fed directly into the bloodstream and regulated carbohydrate metabolism. In November, he took his idea to a leading authority in the field, Professor John J. R. Macleod, at the University of Toronto. Banting himself had no published articles and no research experience, but Macleod agreed to provide him with laboratory space and a student assistant — Charles Best, who got the job by winning a coin flip with another student.

Banting and Best, under Macleod's supervision, began their work in May 1921 with little more than a collection of dogs, a theory, and a prayer. It was messy, bloody, difficult work, sometimes done in sweltering conditions: removing pancreases from dogs, testing their blood and urine for sugar, and observing whether they lived or died. Four died the first week and three more the second, forcing the young scientists to buy replacement dogs. They would work through the night, cooking eggs or steaks over a Bunsen burner and singing war songs to pass the time. To produce the extract, they bound the ducts of a dog's pancreas and removed the atrophied organ. Then they chopped it up and placed it in cold Ringer's solution, to prolong the viability of the excised tissue. The mass was then ground up with sand, warmed, and given intravenously to another dog whose pancreas had been removed and was therefore effectively diabetic.

By August, Banting knew the extract could lower a dog's blood sugar, and a biochemist named James Collip was summoned to apply the solution to humans, to somehow extract the active principle and to purify it — a process that he described as "bathroom chemistry." Using alcohol to remove the contaminants, Collip began injecting extract into normal rabbits, causing their blood sugar to fall. As the extract's potency increased, he noticed that the rabbits would become more hungry, agitated, and then convulsive, ultimately descending into hypoglycemic shock. Michael Bliss's description of the rabbits is a harrowing precursor of how insulin, when overdosed or not offset by food, would affect humans:

Their heads snapped back, eyeballs protruding, limbs rigid, they would violently toss themselves from side to side, then collapse into a kind of coma, lying still on their sides and breathing rapidly. The slightest stimulation, such as shaking of the floor, would set them off again. Sometimes lying on its side the animal's limbs would move rapidly, as in running.

On January 11, 1922, a fourteen-year-old diabetic, Leonard Thompson, received the first injection of insulin, "a murky, light brown liquid containing much sediment." The following day, he received two more shots, causing the sugar and ketones in his urine to disappear. In February, six more patients were treated, with favorable results. In March, the researchers published a paper describing the trial, and Macleod made a formal presentation in Washington, D.C., on May 3 — remarkably, only two weeks short of a year from when Banting and Best began their work.

It had become only a matter of time before insulin was discovered. Other researchers were homing in on it, notably the Romanian scientist Nicolas Paulesco, who in 1921 published articles on successful experiments with pancreatic extracts. But the Toronto researchers developed it first, which earned Banting and Macleod the Nobel Prize in 1923. Banting shared his with Best while Macleod shared with Collip. But the discovery alone did not make the treatment available to the estimated one million Americans who had the disease. The researchers could not produce insulin in large quantities, only laboratory batches, and the drug they did make was often contaminated, weak, or ineffective. Recognizing their limitations, the Toronto researchers accepted an offer of collaboration from Eli Lilly, a family-owned pharmaceutical concern in Indianapolis.

Making insulin on a large scale was gory and difficult. The company had to haul in the beef and pork pancreas glands from the slaughterhouses, inspect them, ice them, and grind them up. It also used rabbits to test the drug; more than 100,000 tests were conducted in the last six months of 1922. By April 1923, Eli Lilly was producing more than 180,000 units of insulin a week; it sold almost 60 million units in the entire year. The operation soon moved into a building that could store a million pounds of frozen pancreata.

"Diabetes, Dread Disease, Yields To New Gland Cure," announced the *New York Times* in May of 1923. The headline typified the public's response to insulin; a cure had been found. That any disease could be effectively treated, let alone cured, was amazing. As John M. Barry noted around the time insulin was discovered, the previous 2,500 years had seen virtually no progress in the medical treatment of patients. The first cure of any disease had occurred only in 1891, when a vaccine for diphtheria was developed, and not until antibiotics were discovered in the 1940s was infectious disease controlled. Insulin was a scientific aberration, a true miracle, its ability to revive the dead giving it a portentous, supernatural quality.

In August 1922, rumors spread through Frederick Allen's clinic that he was acquiring a batch of this new drug. As a nurse recalled, "Diabetics who had not been out of bed for weeks began to trail weakly about, clinging to walls and furniture. Big stomachs, skin-and-bone necks, skull-like faces, feeble movements, all ages, both sexes — they looked like an old Flemish painter's depiction of a resurrection after famine. It was a resurrection, a crawling stirring, as of some vague springtime."

The night Allen was expected, his patients were "silent as the bloated ghosts," impatiently waiting, eyes to the ground. After dinner, they finally heard him in the hall, walking with his wife, "her quick tapping pace making a queer rhythm with his." The patients focused on that sound. When he finally appeared at the doorway, "he caught the full beseeching of a hundred pair of eyes. It stopped him dead. Even now I am sure it was minutes before he spoke to them . . . 'I think,' he said. 'I think we have something for you.'"

Photographs in the *Journal of the American Medical Association* showed the effect of the new drug. Before insulin, a naked fifteen-pound three-year-old boy clings to his mother. His face is grimacing, his ribs are exposed, and his arms are as brittle as twigs. After taking insulin for only three months, a head shot shows the boy with full cheeks, alert brown eyes, and dark locks of hair that fall across his forehead. He looks normal — and cured.

"A new race of diabetics has come upon the scene," Elliott Joslin wrote within a year of using insulin. He called these patients "the erstwhile dead." When one of his patients saw insulin's impact on children, he said, "Now they make noise."

Some doctors had reservations; the therapy itself was unprecedented. Patients, regardless of their disease, had never given themselves injections, so diabetics could endanger themselves by taking the wrong dose. Indeed, some states forbade individuals from even having a hypodermic syringe; in Massachusetts, the prohibition lasted at least until the 1950s. Patients required a doctor's permission to break the law.

Nonetheless, the drug's benefits easily transcended these concerns. Insulin did more than cure. It also redeemed, or — as Allen said — a patient received "salvation through insulin." One such patient, Samuel E. Beckett, wrote to Joslin in November 1922: "If the true Christian be the close imitator of Christ, then the discoverer, and the one who applies the discovery, should feel that they are literally following the one who said: 'I am come that they might have life, and that they might have it more abundantly.'"

Joslin himself, shaped by his Puritan heritage, had witnessed so many "near resurrections" that he compared them to Ezekiel's vision of the valley of the dry bones (Ezekiel 37: 2–10):

> . . . and behold, there were many in the open valley; and, lo, they were dry.
>
> And he said unto me, Son of Man, can these bones live?
>
> And . . . lo, the sinews and the flesh came upon them and the skin covered them above: but there was no breath in them.
>
> Then said He unto me, "Prophesy unto the wind, prophesy, Son of Man, and say to the wind, Thus saith the Lord God: 'Come from the four winds, O breath, and breathe upon these slain, that they may live.'
>
> So I apologized as he commanded me, and the breath came into them, and they lived, and stood up upon their feet, an exceeding great army.

To humanize insulin, however, newspapers did not need to write about the "great army" of diabetics. It needed only one, and it had the perfect one in Elizabeth Evans Hughes.

Responding to her mother's pleas, Fred Banting agreed to treat Elizabeth, who arrived in Toronto on August 15, 1922, and lived in an apartment with her nurse. Almost fifteen years old, she weighed forty-five pounds and had dry skin, brittle hair, and "wasted" muscles. "She was scarcely able to walk on account of weakness," Banting wrote.

She began receiving insulin immediately — two injections a day — which cleared the sugar from her urine. Banting raised her diet from 889 calories a day to 1,200 after one week and then to 2,400 by the second; it was the diet of a normal girl. She ate white bread, corn, macaroni and cheese — all foods that were previously forbidden. But Banting was simply using common sense. If your patient is undernourished, you feed her. He was amused by Elizabeth's reaction, writing: "When I said she was to have bread and potato, both patient and nurse thought that I was joking and breaking faith with the gods, Joslin and Allen. The child was so delighted that she could hardly wait until the next day." The most amazing part of her diet, she told her mother, was she could even eat candy. "Now don't be shocked by that statement," she wrote, "for it is only [for low blood sugars] when I have that privilege." She learned the guilty pleasure of hypoglycemia. "I haven't had a reaction for three days now and I'm wanting to get my piece of candy again like a naughty child," she wrote. "I hope I'll have one when you're up here, so you can see how it affects me."

Her diet still limited overall carbohydrates and was compensated with fats: a daily pint of thick cream and butter balls, prompting Elizabeth to ask her mom in one letter: "What's the sense of taking so much butter when you have no bread to eat it on." But she realized that the insulin was allowing her to recover, to regain her strength, and to fill out her body. In five weeks, she gained about ten pounds; in eight weeks, sixteen pounds. Her watch needed extra links. She grew taller. "I simply don't recognize myself as the same person when I look in the mirror," she wrote in October. A bee sting allowed her to observe: "There are some misfortunes about getting fat and juicy again."

She had the strength to go to movies, concerts, trips to Niagara Falls ("the hills and valleys are literally died (sic) red with the radiancy of the brilliant foliage"), even outings with Banting, who took her to the lab where insulin was being made. She had visions of returning home, living with her parents, attending school. "That is the most thrilling thought to me," she wrote, "to think that I'll be living a normal, healthy existence is beyond all comprehension . . . I declare you'd think it was a fairy tale."

Insulin, of course, made it all possible, but the therapy took its toll. Each lot of insulin varied in potency, making it difficult to find the right dose and causing frequent lows, sometimes three a day. Elizabeth's nurse would have to help her when she bottomed out while sleeping. She wrote: "I'd wake myself up soaking with perspiration . . . having a rapid pulse . . . Blanche would give me half a glass of orange juice and half a glass of cream."

If Elizabeth spilled sugar in her urine, she knew she had weak insulin. Because the glass syringe held only two cubic centimeters, the weak batches would sometimes require Elizabeth to double her volume, beyond the capacity of the syringe. Her nurse would insert the needle and inject the insulin; then, without removing the needle, she would unscrew the syringe, fill it with more insulin, reattach it, and then deliver the insulin. "I feel like a pincushion," Elizabeth wrote. "It really is quite a process, and altogether takes about twenty minutes for the whole performance."

It appears that her injections were given repeatedly in her hip, perhaps because that was the fleshiest part of her body, but repetition, combined with the impurities of the insulin, caused tremendous swelling and pain. "My hip feels as if it would burst, but it doesn't, although my whole leg is numb until I walk on it a bit," she wrote.

But these were minor sacrifices in light of her new condition. Elizabeth empathized with her father when she learned that he had hurt his hip in a fall; her own hips were covered with lumps. "But cheer up it's all in a good cause," she wrote. "I could put up with absolutely anything to keep on this wonderful diet I'm on now." In November, when she heard that an eight-year-old girl had given herself a shot, she demanded to give herself hers as well.

Banting was thrilled, not only in seeing one of his first patients thrive but also knowing the interest in this particular patient. He invited other doctors from Canada and the United States, including Frederick Allen and Elliott Joslin, to visit Elizabeth, who still held Allen in contempt. "He's grown very fat but his nose hasn't filled out any and it's as flat as ever," she wrote. But she adored Joslin, saying he was "the sweetest man. All he could do was to look over at me and smile and say he had never seen anyone with Diabetes look so well."

Her progress attracted the attention of newspapers across North America, which could now put a human face on a medical miracle. Her age evoked sympathy; her famous father — now the secretary of state — made her story even more compelling. "Hughes' Daughter 'Cured' of Diabetes" declared one unidentified newspaper. Elizabeth received many letters, including some from other diabetics who did not have the same access she had to insulin. "The poor things," she said in response to two letters. "I feel so sorry for them."

But Elizabeth disdained the spotlight. "Please don't let on to a newspaper reporter!" she wrote to her mother. "Haven't they been horrible though. I hate to be written up like that all over the country and I think it's cheapening to the discovery. Poor Dr. Banting's even gotten to the point where doctors are beginning to kid him about advertising his discovery through me."

Back home, Elizabeth found that President Warren Harding's campaign slogan — "A return to normalcy" — reflected her own mood. Determined that diabetes would not interfere with her life, she continued her schooling, traveled to Europe in the summer, played basketball, swam, attended Brooklyn Dodger games, and participated in student government. She also smoked and drank alcohol. Along the way, she discovered an unfortunate but common side effect of insulin therapy — weight gain. Uncontrolled diabetics flush out nutrients, but patients who stabilize their blood sugar can retain their food and need to adjust their diets accordingly. By 1926, Elizabeth weighed 158 pounds, more than triple her size since her first injection. She acknowledged that her weight was "too much," and she trimmed down by the time she graduated from Barnard College in 1929. She continued to watch her diet and test her urine several times a week and often acted as her own doctor, adjusting her insulin doses on her own.

Elizabeth's idea of normal did not just refer to lifestyle; it also meant hiding the disease, taking extraordinary measures to conceal the very ailment for which she was celebrated. In 1930, she married

William T. Gossett, a young attorney in her father's law firm, but she didn't tell him about her diabetes until after they were engaged. A two-volume authorized biography of Charles Evans Hughes was published in 1951, three years after Hughes's death; Elizabeth and the rest of the family agreed to cooperate on one condition: the author could not mention her diabetes. In fact, Elizabeth expunged any reference to her disease from her father's papers. She also destroyed any photographs taken during her illness.

She and her husband moved to Michigan, where he worked in the legal department of the Ford Motor Company. Elizabeth raised three healthy children, all born by caesarian section — itself an achievement. (Most diabetic women were discouraged from bearing children.) Reserved and dignified, she was conscious of her pedigree, and her choice of volunteer organizations — Planned Parenthood, the Red Cross, and the Junior League — reflected her belief in civic involvement and public service. She was the very picture of good health, walking regularly, riding her bicycle to work, planting a vegetable garden, and traveling abroad with her family.

Discipline and order were emphasized in the house, but some routines puzzled the children. Each evening at five-thirty, their mother would go into her bedroom, lock the door, and not return for several minutes. She never said what she did, and the children never asked. The family would then gather in the library, and at six forty-five the maid would announce that dinner was served. One evening, however, when the maid entered the library, Elizabeth wasn't there. Her thirteen-year-old son, Tom, went to get his mother from her bedroom. This time the door wasn't locked — a sign that something was amiss — and when he opened it, she was lying on the floor, moaning. He had no idea what was wrong. He tried to shake her, telling her it was time for dinner, but she didn't respond. Just then his father drove up, and Tom raced to meet him at the garage door. "Something's wrong with Mother!" he yelled. "I can't wake her up!"

A stricken look crossed his father's face. He threw down his briefcase, ordered the butler to get orange juice, and ran to the bedroom. He also told the children to go to their rooms, where they waited until they were called for dinner. At the table, their mother looked a bit shaky, but nothing was said of the incident.

Elizabeth concealed her diabetes from her own children, hiding her insulin bottles, syringes, and testing supplies and never allowing them to watch her injections, which she took twice a day. The subterfuge created apparent contradictions: she forbade cookies and other treats in the house, privately fearing they would increase the risk of diabetes in her children, but she herself kept hard candy in her purse. Their paternal grandmother occasionally relieved the children's glucose deprivation by smuggling in sweets. Her husband served as president of the American Bar Association, which meant the couple attended many formal dinners. Elizabeth carried her insulin and syringes in a little bag and would move the dessert around on her plate; at home, she ate grapefruit after meals.

She finally revealed her disease to each child when they were eighteen, taking them out to dinner to break the news and swearing them to secrecy. Why she waited is unclear; presumably she thought they would be mature enough to understand the disease without fear or prejudice. Or perhaps she thought they would be old enough to keep a secret. For their part, the children didn't think any less of their mother. "I wasn't shocked or mortified, because we had such a normal childhood," said one of her daughters, Tony Denning (named after her maternal grandmother, Antoinette). If anything, the children respected their mother even more, for her resilience and courage, and they felt sympathy for her private anguish. "I so admired her for being able to cope with this," Tom said, "but it wasn't until later that I thought it was too bad she had to" maintain secrecy.

The passing of the years didn't change her attitude; she kept the secret from her grandchildren as well. When someone recognized her name and asked if she was that young girl who had been saved by insulin many years earlier, she said no — that was my sister, Helen, who's now dead.

Her children suspect she simply wanted to live a "normal life" and not be treated any differently. It's also likely that she wanted to purge the memories of her starvation years — "like leaving a nightmare," she once said. "In order to put it out of mind, I had to sever all connection with it." She may also have been embarrassed, fearing that others would see her as diminished and concerned that her illness would reflect badly on her famous name or beloved family.

Had she made a different choice, Elizabeth Evans Hughes could have been a role model and a heroine, insulin's original poster girl who symbolized the drug's immense possibilities. But by choosing secrecy, she was truly the quintessential diabetic — the poster girl for misplaced diabetic shame. Her desire for anonymity is reflected in the countless ways in which diabetics camouflage their condition, furtively taking their shots beneath restaurant menus, sneaking glucose tablets when their blood sugar falls, and refusing to disclose their illness to employers, friends, lovers. They have the best intentions; they want to be accepted for who they are, so they do what Elizabeth did: they lie.

In 1980, Michael Bliss was researching his book on insulin. After studying Elizabeth's medical records at the University of Toronto, he located her husband in Birmingham, Michigan, and wrote to him, essentially asking when Elizabeth had died and about the later course of her disease. He soon received a letter from William Gossett's home — written by Elizabeth Hughes Gossett. Now seventy-three years old, she had not slowed down at all, and she agreed to see Bliss after she and her husband returned from a trip to China.

Elizabeth was keenly interested in Bliss's project, wanting to know particularly about the men who had saved her so long ago. She admitted to having fallen a little bit in love with the handsome Charley Best, and she was still grateful to Fred Banting for his liberal diet. She impressed Bliss with her intelligence, her verbal skills, and her appearance: she was a slim, attractive woman who had avoided virtually all the complications of diabetes. Her vision, her kidneys, and her legs were all in good shape. She even proved to be a quick healer when she broke her hip in 1968. She had quit smoking long ago but still drank a cocktail or two before dinner. Surviving that long and that well with diabetes — she had taken some 43,000 injections — requires some genetic luck; but her iron will, her positive attitude, and her stubborn self-reliance also helped. "I don't think too much of doctors or the medical profession," she said. "They don't have enough vision."

Even in the twilight of her life, she insisted that her diabetes be kept secret. She agreed to cooperate with Bliss on the condition that he not use her real name, so he planned to call her "Katharine Lonsdale, the diabetic daughter of a prominent American political figure." When Bliss and the Gossetts went to lunch at the Bloomington Hills Country Club, all discussion of diabetes or insulin had to stop. It would be dreadful, she said, if her friends started giving her advice on what to eat.

Five months later, after a second trip to China, Elizabeth fell ill and was taken to the hospital with a blood sugar of 700. The doctors initially thought she had pneumonia but later discovered that she had suffered, fittingly enough, "silent heart attacks." She died on April 25, 1981, but not before she informed a nurse: "I've had the most wonderful life. I'll tell you about it sometime."

CHAPTER 3

The Burden of Control

AS A STUDENT at Harvard Medical School in 1893, Elliott Joslin saw his first diabetic patient. In the years to come, his name became synonymous with the care of diabetes, his clinic in Boston achieving international renown. He wrote groundbreaking textbooks, expanded the educational role of nurses and nutritionists, lectured to thousands of students and patients, spearheaded summer camps for diabetic children, and helped train two generations of clinicians. By the time he died in 1962, his clinic had treated 58,000 patients, with Joslin himself seeing as many as half of them.

But his approach to the disease was also the flashpoint of bitter debate — a medical brawl that spanned six decades and shaped the treatment, for better or worse, of virtually all patients. Joslin was ultimately vindicated, cementing his stature as a visionary clinician, prescient in his understanding of the disease, but not appeasing his critics. His attitude suggests a more complex legacy. While Joslin could be compassionate and self-effacing, he was also moralistic and unreasonably judgmental. He blamed patients for bad medical outcomes, accused them of betrayal for unintended lapses, and reinforced the guilt that can demoralize patients who understand that their best intentions cannot always control an exasperating condition.

Elliott Joslin created an abstemious ideal for the diabetic, one that grew out of his own upbringing in Oxford, Massachusetts, about fifty miles west of Boston. He had deep roots in the region's Puritanical soil; according to Joslin, an ancestor was sentenced to death during the Salem witch trials, only to be spared by the English governor.

(When Joslin was awarded an honor in England, he thanked the British Crown for saving his life.) Heavily influenced by the Congregational Church, Joslin hung in his office a photograph of John Wesley, the Anglican clergyman who was a founder of Methodism. Wesley opposed predestination, believing that each man could control his own fate through faith in God and virtuous conduct — a message of personal responsibility that Joslin would impart to his patients.

Joslin knew the grim realities of diabetes. It killed his aunt in 1898, and early in his career he watched three diabetic women in his hometown "hopelessly suffer and wither away." Death, in fact, came with staggering consistency. In 1956, Joslin noted that 15,795 of his patients had died. Nonetheless, even before insulin, Joslin was curiously sanguine about the diabetic's plight. In his *Diabetes Manual* (1919), he wrote: "QUESTION: What can a diabetic patient do for himself besides keeping his urine sugar-free? ANS: Be cheerful and also be thankful that his disease is not of a hopeless character, but a disease which his brains will help him conquer." Diabetes, in fact, was "of a hopeless character" in 1919, but Joslin's optimism set a tone for unrealistic expectations that would accompany diabetes research for the next eight decades. His message also reflected his view that the illness was not strictly a metabolic disorder but a profound moral challenge that tested the character of its patients. Those who mastered the disease, or at least those who delayed its fatal course, did so through superior strength and intelligence. The "honest diabetic" would be a focus of Joslin's long career. He told the Johns Hopkins Medical Society in 1918: "This is a disease which tests the character of the patient, and for successfully withstanding it, in addition to wisdom, it demands of the individual honesty, self-control, and courage."

The person who perfectly embodied these ideals was his mother, who was diagnosed with diabetes in 1899, becoming his eighth patient. Overweight, she almost certainly suffered from type 2 diabetes, which allowed her to live — with the help of a strict low-carbohydrate diet — for fourteen more years, dying at the age of seventy-three. That was long enough to earn a medal from her son for "Exceeding her Life Expectancy." As he later recalled, "She never broke the Naunyn diet, never mentioned the disease, resented sympathy, lived bravely, [and] carried on a whole social service of her own for the old and the decrepit, rescuing alcoholics in my father's factory."

Joslin always put his patients at the forefront of their own care, writing in his manual that "the patient is his own nurse, doctor's assistant and chemist." While he praised those who successfully managed their disease, he faulted others for their own demise. In the spring of 1922, for example, he said that gangrene, coma, and infection could all be prevented, so when a patient died in coma, it could be "traced to the advice of the laity or irregular practitioners, and often to the patients themselves." Insulin more clearly placed diabetics in control, transferring "responsibility for the maintenance of the life of the diabetic from the doctor largely to the diabetic himself." The medication, he added, has "lightened the load on the doctor's back and given the patient the means of completing his life's journey."

The first ten or fifteen years after the discovery of insulin was a period of naïve expectations — based on the belief that diabetes had been solved, if not necessarily cured. Joslin, however, feared that the miracle drug would erode patients' self-control, tempting them to eat at will and destroying all nutritional discipline. He believed the product had to be used with diet and exercise, advocating low doses to deter patients from the lure of "overnutrition." He wrote in 1923, "We do not believe in diabetic patients walking on insulin stilts any more than is necessary."

By the 1930s, insulin's halo had begun to tarnish, for it became clear that patients who had the illness for at least ten years were prone to problems of the small blood vessel, specifically in the eyes and kidneys. Other complications, including coronary disease and impaired nerves, also developed at high rates, and the fear arose that blood vessel damage extended through the entire body. The era of complications had begun.

This development fortified Joslin's insistence on rigidly managing the disease, and he argued that "tight control" would minimize these risks and increase life expectancy. Control, however, was measured crudely and imperfectly through urine tests. Each required collecting a fresh sample and mixing eight to ten drops in a test tube with Benedict's solution; named after an American chemist, it would change color in the presence of glucose. The liquid boiled for three minutes — "cooking urine," as some called it. The resulting color indicated sugar content: blue signified no sugar and shades of green, yellow, and red reflected increasing concentrations. (By the middle 1950s, simpler methods had been developed, including self-heating tablets and test strips.) The tests were time-consuming, cumbersome, and, most important, a poor substitute for blood sugar levels, in part because the urine may have been in the bladder for hours and was therefore "stale." But these tests were the best measure of control during Joslin's career.

Test results inevitably fluctuated, particularly before the development of more physiologic insulins. But Joslin compared a urine test to a math test: success required hard work and should be rewarded; failure was a sign of laxity and should be stigmatized. He said in 1958, "The idea has been raised that a diabetic should not feel he has done wrong if he has a poor Benedict test. I disagree. When I see a red test, I know, if uncorrected, that patient is headed for destruction. I believe he should be happy when he knows his faithful following of the rules ends with a blue victory." Joslin exalted the superior virtues of patients who maintained tight control, and in 1946 he spelled out his Diabetic Creed: "That adherence to treatment with diet, insulin and exercise finds ample justification in the good health, comfort and longevity of those who obey the rules as contrasted with the poor health, suffering and shortened lives of those who are careless." Joslin rewarded faithful diabetics by introducing the Victory Medal in 1947 for those who had had the disease for at least twenty-five years and were found to be free of complications. Several years later, those who had outlived their normal life expectancy were honored with another medal, tellingly inscribed: "A Scientific and Moral Victory."

Joslin didn't necessarily trust his patients, preferring that they eat oranges at his clinic instead of other fruit so that nurses could track the peelings. He refused to acknowledge the discomfort of injections. Until the late 1930s, patients used large-bore 25-gauge steel needles, which were often burred and dull and looked like miniature harpoons to any child. But in 1923 Joslin said, "The pain of insulin injection is slight. Otherwise children would not so readily give it to themselves."* Joslin also believed that diabetics who fell short in their control were indifferent or indulgent; obese patients were scorned. Though the distinction between autoimmune and insulin-resistant diabetes had not yet been recognized — type 1 and type 2 — the disease had already been connected to obesity. In 1921, 40 percent of Joslin's patients were overweight, prompting him to write that diabetes was "largely a penalty of obesity, and the greater the obesity, the more likely Nature is to enforce it." His critical view of fat people is a bit surprising, given that his mother was hefty. Nonetheless, while acknowledging that even those who eat moderately can gain weight, he compared the obese diabetic to the alcoholic — the latter is viewed with pity or contempt, and perhaps the former should be viewed that way as well. "In the next generation, one may be almost ashamed to have diabetes," Joslin wrote.

He was even more critical of patients who had hypoglycemic reactions. With symptoms of trembling, perspiration, and confusion, low blood sugars inevitably occur in anyone who tightly manages the disease. But in an undated speech, Joslin compared those who had such lapses with turncoats, saying, "The diabetic who allows himself to have a reaction in public is a diabetic fifth columnist, because by doing so he injures the reputation of every other diabetic soldier in the army." While hypoglycemia could contribute to discrimination against diabetics, Joslin's intemperate view only intensified the feeling of regret of any struggling patient. In his own way, he legitimized the shame, even self-loathing, that diabetics felt in his era and persists to this day. As noted by Chris Feudtner in his book *Bittersweet,* Joslin's commitment to control "was entrenched so deeply that it became a sacred ideal, beyond criticism," so that when medical problems arose, a patient "carried the burden not simply of

---

* In fact, children were rarely eager, even when the needles were smaller and less painful. In the 1960s, when the Joslin Clinic began changing patients to multiple daily injections, the children and their parents angrily objected; one kid "literally fled the room and turned somersaults down the unit's corridor," according to Dr. Barnett.

his failing health but also of the medical attitudes that construed complications as the result of poor patient compliance."

Joslin's contemporary critics made similar claims, saying that he strapped patients in a medical straitjacket, injuring them emotionally while increasing their risk of hypoglycemia with his emphasis on normal blood sugars. They also asserted that the benefits of tight control were scientifically unproven and advocated instead an alternative. Aptly named the "free diet," it could have been called the "anti-Joslin." Its chief proponent was certainly a nemesis.

Edward Tolstoi was Joslin's opposite philosophically as well as culturally. Joslin, the reserved Boston Brahmin, preached denial and self-control; Tolstoi, the extroverted New York immigrant, wanted to liberate diabetics from their culinary prison. By 1949, Tolstoi had been practicing for twelve years and had seen more than three thousand patients. While he didn't invent the free diet, he promoted it through books and lectures, often crossing paths with Joslin. In 1940, Tolstoi was a founding member of the American Diabetes Association; Joslin was its first honorary president. Their debates were so memorable that they would be reenacted for physicians decades later.

Tolstoi argued that Joslin's diet, which involved weighing food at each meal to calculate calories, was burdensome, impractical, and embarrassing. "It sometimes seems the diabetic devotes more time to worrying about his meals than to eating them," he wrote in 1952. "It is always so many grams of protein, such and such percentage of vegetables, and so many ounces of juice, milk, etc. It isn't food. It's mathematics." He scoffed at the goal of keeping the urine free of sugar, believing that patients act on natural instincts to eat but are then accused of "dietetic sins." "The result, for the patient, is despondency . . . as well as frustration and guilt because he is led to believe that he made it so by not adhering to his diet." Because the clinical benefits of tight control had never been proven, Tolstoi said, its emotional burdens were all the more reprehensible. That rigid management also increased the risk of potentially life-threatening hypoglycemia made it even more odious to him.

The free diet approach encouraged patients to "satisfy [their] hunger," eating "sweets in reasonable amounts" and allowing them "to share as comfortably as possible in the normal enjoyment of living." Sugar in the urine was disregarded. The goal was to keep patients free of "symptoms," meaning excessive thirst or hunger, weight loss, or acidosis. When these occurred, the diabetic could return to proper balance by increasing the insulin dose. Tolstoi recognized that many diabetics had blood vessel damage, but he concluded that patients in tight control had the same rate of complications as those who followed the free diet. "This does not paint too bright a picture for the diabetic," he conceded, "but some of the pessimism is nullified by the equally true observation that not *all* diabetics develop these changes in the blood vessels." So why do some diabetics experience complications and others don't? It was, he wrote, "a mystery to science."

What Tolstoi neglected to mention was that his approach all but exonerated physicians from responsibility. If complications were inevitable, then doctors need not worry about the quality of their care. If tight control had negligible benefits, then they need not spend the time and resources to educate and train their patients. It is not surprising that Tolstoi's approach was followed by so many general practitioners: the free diet gave them a free pass.

The central question debated by Joslin and Tolstoi — were diabetic complications a genetic inevitability or could they be prevented through rigorous management? — created a fissure in diabetes care for much of the twentieth century. Joslin's view, known as "the glucose hypothesis," appealed to common sense: of course diabetics should strive for normal blood sugars; aberrations by definition are unnatural. Joslin's problem was that he had no way of testing his theory, no means of controlling his patients' blood sugars so they could approximate those of nondiabetics, and no tool to measure the actual value of control over time. Tolstoi was correct in saying that patients who tried to stay under tight control had a high incidence of complications, but that's because, given the available therapies, normal glycemia was rarely achieved. Tight control didn't exist; just gradations of poor control.

By the 1980s, however, several important technologies had been developed that could finally resolve the controversy. The first, the home glucose meter, proved to be the most important breakthrough since the discovery of insulin. The meter's precursor was a test strip, designed by the Ames Company in the 1960s. Patients could prick their finger, put a drop of blood on the strip, wait sixty seconds, wash it off, and tease out glucose levels by determining its color. Primarily used in doctors' offices, the strips could measure only extreme highs or lows, so they were of little use. By the early 1970s, Ames used the same strips for a monitor that indicated glucose levels through a swinging needle. This too was intended for doctors' offices, as the high cost (about $500), bulkiness, and imprecise readings made it impractical for home use. Only by the late 1970s did self-monitoring become a focus, in part because endocrinologists who were caring for pregnant women believed that a normal blood sugar gave their patients the best chance of delivering healthy babies. By the early 1980s, several meters targeted for patients — smaller, portable, cheaper — had been introduced; and in 1986 the first biosensor meter, which improved accuracy by reading glucose levels with an electric current, became available. A multibillion-dollar market was born.

Patients could now get current glucose readings, but a true measure of control could not be established until a second test was developed — the hemoglobin A1c, which indicates average blood sugar levels from the past several months. Hemoglobin is part of the red blood cell that carries oxygen to the tissues and organs. It mixes with blood glucose, which attaches to the cell and makes hemoglobin A1c. Because a cell's lifespan is 90 to 120 days, the A1c reflects the concentration of glucose in the blood during that time, though it is more heavily weighted toward the last thirty days. Once A1c levels could be quantified, the values became the gold standard of control — a "truth test" — and patients could no longer deceive their doctors or themselves about their control over a period of time.

The glucose meter and the A1c test set the stage for one of the most ambitious clinical trials in medical history. Starting in 1984, the Diabetes Control and Complications Trial (DCCT) involved 1,441 volunteers with insulin-dependent diabetes at twenty-nine medical centers across the United States and Canada. Half the patients followed intensive management therapy designed to keep their blood sugar in the near-normal range. That meant frequent daily blood tests, close attention to diet and exercise, and at least three injections a day, or their equivalent, through insulin pumps or syringes. Patient education, with assistance from nutritionists, ophthalmologists, and other specialists, was also stressed. The results were significant. The experimental group's average A1c was 7 percent, or a blood sugar of about 170 mg/dl, while the control group, with no more than two daily shots, had an average A1c of 9 percent, or about 240 mg/dl. With improved methods to measure eye, kidney, and nerve disease, the researchers could now evaluate the extent to which tight control affected complications.

The first few years saw little divergence between the two groups, but over time clear differences emerged. In fact, the study was supposed to last ten years, but it ended a year early because the results were so convincing. Intensive therapy patients saw their risk of developing retinopathy reduced by 76 percent; progression of the disease was slowed by 54 percent. The incidence of neuropathy, or nerve damage, was reduced by 60 percent. A follow-up study of six additional years, with the same patients, showed that tight control reduced heart attacks and strokes by 57 percent — no small discovery, given that cardiovascular disease is the number one cause of death for all diabetics.

Tight control, however, was not without its own risks. In the original study, it tripled the rate of hypoglycemia; the intensive group also found that severe low blood sugars, resulting in coma or seizure, were relatively rare but still three times as likely.

Those risks did not dampen the enthusiasm over the results. The trial itself, which cost $200 million, was celebrated for its length, commitment, and thoroughness. It covered almost 10,000 "patient years," with only eight volunteers dropping out; one investigator, David Nathan, said the study "had the best adherence in the history of science." More important, the results proved that diabetic complications were not inevitable, that rigorous control did matter, and that Elliott Joslin had been right. Thirty-one years after his death, his vindication was complete. After the results were announced, the Joslin Clinic passed out buttons that read: "I Told You So."

But if Joslin's argument on blood sugar was vindicated, so too was his unforgiving view on individual responsibility. If tight control can stave off complications — if diabetics have the right tools to achieve their normal life expectancy — then any failure will lie with the patient, not with the disease itself. Blindness, amputations, kidney failure, arteriosclerosis, neuropathy: the rogues' gallery of diabetic plagues can now be traced to the individual's own mismanagement — or, Joslin would argue, his moral failures. But some physicians, including those at the Joslin Clinic, believe that finger-pointing worsens outcomes.

"The problem with the DCCT was the whole fixation on causality, and the other side of causality is blame: 'It's your fault,'" said Howard Wolpert, a senior physician at Joslin. Some of his patients have been badgered by other doctors for failing to meet glycemic targets, and he refers to some physicians as "fascists" in their use of high-pressure, accusatory tactics. Wolpert, whose empathy for patients has been influenced by his diabetic wife, notes that even the language of care — blood sugar *tests,* A1c *tests,* ketone *tests* — implies judgments about character, discipline, and morality. When patients believe they have "failed" these tests, their frustration can lead to a downward spiral of despair or apathy.

Moreover, the DCCT created unrealistic expectations. Its volunteers, already highly motivated, had access to teams of specialists at academic centers and were never limited by financial considerations, circumstances that few patients can duplicate. Those following intensive therapy "were assigned nurse practitioners who hounded them and cajoled them and brought them in constantly and literally bribed them with theater tickets and baseball tickets," said Richard Kahn, the chief scientific and medical officer of the American Diabetes Association. In fact, the original A1c goal for the intensive group was 6.05 percent, but that proved unrealistic. In subsequent studies of the same volunteers — in which they did not have comparable support — the average A1c increased to 8.0 from 7.0. "The first lesson," Kahn said, "is that it's very difficult to achieve these goals."

Nathan said participants were invited to an annual social event, usually a dinner, a cookout, or a ball game; "but given the efforts and close relationships that we forged, that level of interaction could not be construed as a 'bribe.'" The trial's success reflected the intensity of care and the innovative use of different insulins to fit the schedules and lifestyles of the volunteers. "In current practice," he said, "most clinicians give up trying to achieve goal A1c's after they have been frustrated, often blaming the patient. In the DCCT, we refused to give up on anyone and continued to try to discover the barriers" to those goals.

Suppose the DCCT's results had concluded the exact opposite — that tight control made no difference and that normal blood sugars were immaterial. That would have been depressing but liberating. The grind of intensive therapy would have been replaced by fatalistic exuberance. Tolstoi was right: cheesecake for everyone! We'll die when we die and we can't do anything about it. That attitude would remove all diabetic guilt. Guilt over eating too much. Guilt over high blood sugars. Guilt over high A1c's. Guilt over the first microaneurysm on your retina, over the first trace of albumin in your urine, over the first tingling sensation in your limbs. There would be no more guilt over your own fallibilities, no more guilt over simply feeling guilty.

That day, of course, will never come, but doctors need to recognize the heavy demands of rigorous management — financial, emotional, and physical — while acknowledging that misjudgments — sometimes small, sometimes large — can never be eliminated completely. Deb Butterfield was a recruiter on Wall Street when neuropathy resulted in the amputation of half of her toe. She later wrote: "Certainly control is important . . . [but] the expectations that have been propagated by the DCCT obscure the truth that diabetes is the problem, not the people who have it."

CHAPTER 4

The Diabetes Queen

FEW PEOPLE COULD HAVE FORESEEN that diabetes would become a mainstream disease, let alone America's biggest epidemic. The opposite was true: insulin's discovery raised expectations that the illness would soon be eliminated. Instead, insulin transformed diabetes into a chronic condition, and the development of oral medications in the 1950s gave a different category of diabetics a less intrusive therapy. Lifespans improved, which gradually increased the number of patients, but only in the 1990s did diabetes emerge as a public health crisis, disproportionately affecting groups that already had diminished access to medical services. These patients, often low-income minorities, confronted a diabetic ethos of self-care — indeed, a therapeutic philosophy that seemed to blame patients for their own failures — but they often lacked the resources to rescue themselves. They would have to find other means, in places like Georgetown County, South Carolina, where profound lifestyle changes had unleashed the diabetic tide.

Florene Linnen was raised in Georgetown, a coastal community whose most famous product — rice — was once known as "Carolina Gold." In the 1840s, the county exported more rice than any port in the world, and Linnen's ancestors, as slaves, contributed to the plantation owners' riches. Agriculture was the county's lifeblood for another hundred years, and Linnen herself was born on a family farm in 1943. She and her ten siblings toiled in the hot fields, harvesting cotton, tobacco, and vegetables, never lacking an incentive to work. For the most part, what they ate was what they grew: collard greens, tomatoes, cabbage, okra, squash, and white potatoes. Only after church or on special occasions did they enjoy a big meal: fried chicken, macaroni, mashed potatoes, ice cream, and apple pie. "Fried chicken was a celebration," she recalls.

Linnen's parents insisted that the children spend time outside, which seemed perfectly sensible. Their house had no air conditioner or even a fan, and certainly no television. The family didn't own a car until the 1960s, so the youngsters walked to school, to church, and to visit their neighbors. They entertained themselves by acting out poems read by their mother or by dancing to the song "Shoo, Shoo, Baby!" played by their father on his guitar.

It was a life of constant motion, which kept everyone lean and fit if not necessarily healthy. Poor medical care meant that any illness could be fatal. Two of Linnen's siblings died as babies. Doctors, clinics, and pharmacies were in town, sixteen miles away — too far for anything but an emergency. Medicine was expensive, so the family relied on home remedies, special teas made from boiled sassafras roots or pine needles, to treat fevers or infections.

Linnen dropped out of school after tenth grade to earn money as a baby-sitter. She soon married and started her own family, in time raising six children. But her kids had a different home life. Consolidation in agriculture drove most family farms out of business. In the 1960s, Linnen's father began working in the Georgetown steel mill, and her husband worked at a lumber company. The Linnens weren't rich, but they had enough money for a car, a television, and an air conditioner. Florene also found a job. She began as a volunteer at the Waccamaw Economic Opportunity Council, helping low-income families pay their heating and cooling bills, then became a paid employee. Working outside the house brought in much-needed money, but it also meant less time to cook. Over time, the family began to take advantage of a completely new kind of eating experience: fast-food restaurants. Quick, inexpensive, and filling, they were considered a godsend.

But these changes — a less active lifestyle, a fattier diet, and a few more amenities — had dramatic implications. When Linnen got married, she weighed ninety-five pounds. Twenty years later, she weighed more than two hundred. She was also tired and had blurred vision, numb hands, and an upset stomach. In 1983, when she was forty, her doctor told her she might have a kidney infection and sent her to the Medical University of South Carolina in Charleston. There she was diagnosed with type 2 diabetes. She knew nothing about it but had few concerns. Many African Americans in Georgetown had "just a little bit of sugar," and no one seemed to worry. "It's not bad at all," she was told. Her doctor, a general practitioner, gave her some pills, and the most severe symptoms disappeared.

Then, in 1997, Linnen attended a diabetes workshop in Myrtle Beach, and a speaker asked, "What were your numbers the last time you saw your doctor?" Linnen had no idea what he was even talking about, and only later she discovered he was referring to her blood sugar, A1c, and cholesterol levels. Even more alarming was the news about diabetic complications: she learned that feet and toes could be amputated because of infection or nerve damage. Linnen was angry. She realized that the medical profession had betrayed not only her but all the diabetics in her community, that modern lifestyles were conspiring against them, and that ignorance was sealing their doom. "Diabetes crept up on us," Linnen says. "We weren't taking it seriously and that's why it overwhelmed us."

She decided to change her lifestyle to avoid being another diabetic casualty. In the process, she took the reins of a quixotic local outreach program that has tried to overcome the many obstacles to care: not just ignorance of the disease but poverty, racism, and deeply ingrained dietary habits. Even religion, a source of strength and comfort, has complicated treatment by causing patients to embrace their faith in lieu of medicine.

Such a challenge has forced Linnen to use unconventional methods to deliver her message, such as showing photographs of feet with four amputated toes or asking an undertaker to give a short speech at a fundraiser. "Either you live right," he said, "or you're coming to see me." Linnen has considered bringing him back, this time with a casket.

About 23,000 African Americans live in Georgetown County, and according to several surveys, about 15 percent, or 3,450, have been diagnosed with diabetes. Another thousand are estimated to have it but have not been diagnosed. Linnen believes that at least half of the black community has diabetes. Whatever the true number, the devastation is undeniable. Black diabetics in Georgetown on average have higher A1c levels, higher blood pressure, and higher cholesterol levels than whites while they receive less diabetes and nutrition education. In some African American families, the disease is ravaging three generations at once, causing victims to undergo amputation, dialysis, laser treatments, and organ transplants. They've burned their feet on hot pavement because they've lost sensation in their nerve endings. They've been disabled, hospitalized, and buried. Surrender is a common response. When Linnen recently visited one woman at her house, she encouraged her to take walks. The woman looked at her blankly. "My dad had an amputation. Now I'll have an amputation," she said. "My mother died of diabetes. Now I'll die of diabetes."

To stop the carnage, Linnen delivers a message that mixes defiance with compassion, anger with self-empowerment, imploring residents to demand more from health care providers. But ultimately she wants her friends and family to save themselves. "I plant the spirit that they are in control," she says. "It's their choice to live or die."

All diabetics walk on a gentle precipice, but for some the path is narrower, the ridge closer, the descent steeper.

That type 2 diabetes disproportionately affects minorities in both incidence and complications is not surprising. Disparities in health caused by poverty, racism, or lifestyles have been well documented. In 2005, the CDC reported that African Americans have higher rates of stroke, high blood pressure, infections, and diabetes, dying from nearly every major disease at rates higher than those of white people. The agency also reported that the number of potential years lost because of strokes, diabetes, and perinatal diseases was three times higher for black people younger than seventy-five than for white people the same age. In 2002, the Institute of Medicine, an independent organization that advises the federal government, reported that members of racial and ethnic minorities were less likely to receive good health care even when they earned as much money as whites and carried the same insurance. David Satcher, a former U.S. surgeon general, estimates that 84,000 African Americans die needlessly each year because of health care disparities. "Racism within the health system is literally making people of color sick," said Will Pitz, a regional organizer for the Northwest Federation of Community Organizations, a health care advocacy group.

The health gap in diabetes, already bad, will only get worse. According to the CDC, in 2002 the age-adjusted prevalence of the disease in non-Hispanic whites was 7.9 percent — compared to 11.7 percent among Hispanic-Latino Americans; 12.9 percent among non-Hispanic blacks; and 17.6 percent among American Indians and Alaska natives. These numbers are closely tied to the growing prevalence of obesity in minority groups, which is the single strongest predictor of type 2 diabetes. According to 2002 data from the CDC, 62 percent of non-Hispanic white adults were overweight — compared to 70 percent of black adults and 73 percent of Mexican American adults. What's more, the rise in childhood obesity — nearly tripling in the past three decades, with more than 15 percent of children between six and nineteen now obese — ensures that the gap between whites and minorities will widen. The epidemic of overweight children has also led to type 2 diabetes in youngsters, which was virtually unheard of in the early 1990s. In fact, type 2 diabetes used to be called "adult-onset," but in 1997 the American Diabetes Association recommended that the term be dropped as misleading.

Even when broken down by racial groups, the cumulative numbers for diabetes fail to convey how it decimates specific communities. Starr County, Texas, for example, sitting on the Mexican border, is 98 percent Mexican American and has only one doctor for every 3,412 residents. Almost half of the adults in the county have type 2 diabetes, and the epidemic is in its infancy. In 2003, a researcher began screening elementary school children, assuming that of the 2,931 children she checked, about 600 would be at high risk for diabetes. She found 1,172. Part of the problem was the free breakfast and lunch served by the school; they were so fattening that they put each child on track to gain nine pounds during the school year. Fifty percent of elementary school boys are now overweight or obese; the number is 35 percent for girls. "I will take you down the hall in any one of my schools, and you will see most of the children aren't slim anymore," said Roel Gonzalez, the superintendent of the Rio Grande City Consolidated Independent School District. "Kids are thirty, forty pounds overweight already, and they're only in high school. We're basically walking time bombs."

The influx of Asian immigrants, especially from Far Eastern nations like China, Korea, and Japan, will also significantly add to the diabetic burden. Confronted with a high-fat Western diet and more sedentary lifestyle, many of these immigrants gain weight. Unfortunately, they also develop type 2 diabetes at lower body masses than people of other races, and, at any size, they are 60 percent more likely to get the disease than whites. The trend could overwhelm New York, forcing a new generation of Asian immigrants to make difficult choices that "could mean the difference between a healthy city and a colony of the sick."

Researchers have developed a theory about the genetic underpinnings of the obesity crisis that suggests minorities are genetically more susceptible to weight gain. According to this hypothesis, evolution favored genes that promoted the accumulation of fat so that humans could survive long periods of famine or natural disasters. Ideal for hunting-and-gathering societies, these "thrifty genes" maximized the amount of energy that could be stored from every calorie consumed. Those societies, however, have mostly given way to industrialization, modernization, and abundance. The result: bodies that were genetically designed to survive starvation, endure physical hardship, and flourish on a low-fat diet are now bloated from affluence, leisure, or an oversupply of fatty, high-calorie foods. When thrifty genes meet an indulgent lifestyle, "diabesity" results.

This theory can explain the general fattening of America; currently more than 45 million adults in the United States are obese, or about 32 percent of the adult population. But even that figure understates the severity of the crisis. Geneticists believe that some groups, particularly those that once had to survive in harsh climates, whose food supply was often scarce, or whose lands were harmed by colonizers, have the highest incidence of the thrifty gene. Achim Gutersohn, a German geneticist, estimates that 90 percent of those of African descent have thrifty genes. Other high-risk groups are Hispanics and American Indians — in other words, the very groups in this country who often receive the worst health care and have the fewest resources to combat the problem.

Type 2 diabetes is also unusually insidious. Type 1 is the equivalent of a head-on collision, with a sudden and dramatic impact. It involves undeniable suffering, but the youngsters who typically develop it are able to adapt. Treatment begins immediately, adjustments are made, life goes on. Type 2 has a very different profile. It is a stealth disease; people can function for years without knowing they have it. Instead of a head-on collision, the car limps along without having to leave the road. But the years of elevated blood sugar take a toll so that by the time the disease is discovered, complications may have already begun. In some cases, patients are diagnosed on the same day they learn they need a toe or foot amputated; that happened to the former professional football player Jack Tatum, who lost his left leg beneath the knee. Even after diagnosis, denial and neglect are common. Most type 2s are adults, with well-established habits, and they may be less willing to diet, exercise, or take medication, including insulin, for a disease whose invisible nature has thus far created the illusion of wellness.

Even when the epidemic is known and chronicled — even when government health officials invest time and money in addressing the problem — it can still overwhelm a community. Nowhere is that truer than with the Pima Indians on the Gila River Reservation in southern Arizona.

In 1963, scientists and doctors from the NIH visited the Pimas to study rheumatoid arthritis. Known as "River People," these Indians had lived in the region for two thousand years, developing an irrigation system to grow wheat, beans, squash, cotton, and other crops. The researchers had an image of the Pimas: muscular men who were hardy runners, fishers, farmers, and hunters; trim women who gathered food and weaved blankets and clothes. But that world no longer existed.

In the late nineteenth century, the Pimas' reliable agricultural system faltered when the white settlers diverted their water, leading to prolonged periods of drought and famine and forcing the Pimas to rely on the lard, sugar, cheese, and white flour provided by the federal government. The twentieth century saw the continued assimilation of the Indians into American life. During World War II, many worked in urban factories, returning to their reservation after the war with significantly more money in the bank. The cash allowed for more leisure time. Physical activity declined. Thanks in part to the government's continued distribution of unhealthy foods, the Pimas' traditional high-fiber diet changed so that fat represented 40 percent of their caloric consumption. Corn tortillas, for example, were replaced with "fry bread," white flour cooked in boiling lard.

While weight gain was seen in many groups after World War II, the NIH researchers were still surprised when they met the Pimas in 1963: half of the adults were morbidly obese and had type 2 diabetes, and 80 percent over the age of fifty-five had the disease. These observations, in fact, led the geneticist James V. Neel to develop the thrifty gene theory.

After the Pimas met the scientists, they agreed to an extraordinary collaboration in which the Indians volunteered for research studies for nearly four decades — it's been called "the most intensive and extensive continuous high-quality scientific investigation on a single homogeneous population, in a localized enclave, in our nation's history." What emerged were new insights into the complex interplay between genes and the environment and the role that physical activity and diet can play in delaying or preventing diabetes. The clinical data should have been used to benefit all people with, or at risk for, type 2 diabetes. But that hasn't happened for any group, including the Pimas. Despite all the research, they are worse off today than they were in 1963, with the highest prevalence of type 2 diabetes in the world.

"Is there an explanation?" asked Arthur Krosnick, a distinguished clinician and researcher who has worked in diabetes for more than fifty years and has studied the Pimas' problem. He noted that they now have improved legal rights, ownership of land, free health care, and free education, and they operate a casino that provides them with a new source of revenue. He was also impressed with a National Diabetes Program run by the Indian Health Service, which is part of the U.S. Department of Health and Human Services. But no government program can return a community to its roots, transform deeply entrenched lifestyles, or reverse a painful history that creates barriers to care. American Indians are justifiably skeptical of the good intentions of Anglo Americans, and cultural or language barriers create other problems. Krosnick, for example, described the Indian mother who ran to the health clinic when her son told her that the nurse said he had diabetes. The mother screamed at the woman for putting a curse on her boy. In another instance, an Indian refused to follow the advice of his doctor because the physician unknowingly insulted the patient by making inappropriate eye contact.

These cultural snafus, combined with limited resources, reinforce the downward spiral. The degenerative nature of complications creates despair, and the omnipresence of the disease leads to fatalism. Under these circumstances, outreach programs for the Pimas have little chance of success. "Noncompliance," Krosnick concluded, "was probably inevitable."

Florene Linnen has a favorite saying: "I ain't never hear the devil had a funeral, so he's still alive." In such a world, only hard work, good intentions, and faith can prevail — an ethos that must be redoubled to save the diabetics of Georgetown County.

Before she attended the seminar in Myrtle Beach, Linnen probably had had better care than many others in the community. She was taking insulin, which is commonly needed in overweight type 2s but is often absent, particularly among poorer patients. Nonetheless, she still suffered from fatigue and blurred vision and was motivated to change her ways. She began taking daily walks and eliminating the foods — fried chicken, macaroni, ham hocks, and pies — that had contributed to her weight gain. She drank water instead of Coke or Pepsi, substituted fat-free milk for whole milk, nixed regular ice cream for sugar-free. She lost twenty-one pounds, and her huge daily dose of insulin fell by 53 percent, from 110 units to 51. She began testing her blood sugar with a home glucose meter and received her first A1c test, and she became an advocate, starting with her mother, who had been diagnosed several years earlier. When she took her mother to the doctor, a nurse measured her blood sugar and reported the results.

"Just a little high," she said.

"How high?" Linnen asked.

"Three hundred eighty-two."

"A little high!" Linnen erupted. "That is not a little high!" She demanded to see the doctor, who would increase her mother's medication.

She makes similar demands on her own complacent physician. "When he doesn't check my foot, I'm now bold enough to stick it up to him," she says

Linnen wanted to broaden her help, so she called a meeting at her church to discuss diabetes. To her surprise, seventy people showed up. She then held a meeting at another Georgetown church, and seventy-five more came out. Most of them knew little about the disease, so Linnen founded the Diabetes CORE Group, which raised money for educational efforts. Her fledgling enterprise received a boost in 1999 when a government program was introduced to reduce racial disparities in diabetes in Georgetown and Charleston counties. The program, called REACH (Racial and Ethnic Approaches to Community Health) 2010, is funded by the CDC and the state of South Carolina, and it tries to raise awareness through seminars, information fairs, and support groups. Linnen's own medical history gives her special standing as an advocate, and the program hired her for $27,000 a year, plus health benefits, as one of five community health advisers.

She takes her campaign to basketball games, church services, clinics, beauty parlors, and neighborhood parties, distributing flyers that outline the medical questions diabetics should ask and the blood sugar goals and lab results they should strive for. She uses a combination of entertainment and shock to convey her message. She has asked beauticians, for example, to fix the hair of women for a program called Diabetics Looking Good. "We get them fixed up," Linnen says. "The guys get dressed in black suits, and those ladies are so happy. Just because you're diabetic, you can still have a social life."

She also shows photographs — under the headline "The Consequences of Diabetes" — of diseased feet, kidneys, arteries, retinas, and gums. She displays tubes of salt, grease, and sugar that indicate what goes in your body when you eat potato chips, hot dogs, hamburgers, or ice cream. To explain food portions, she lays out plastic samples — the kind used in a children's kitchen set — of chicken, cornbread, green beans, and pears. All food, she tells her audiences, should be weighed or measured, and if you don't have a scale or cup, use your hand: a piece of cornbread should not be larger than your palm.

With straight, jet-black hair, broad shoulders, and a robust laugh,

Linnen has appeared on radio and television and has given her lectures in New Orleans, Atlanta, and elsewhere. Her passion has earned her a fitting nickname: "Diabetes Queen." But her relentlessness can also be overbearing. She once approached an overweight shopper who was buying unhealthy foods. "Are you diabetic?" she demanded.

Her daughter-in-law, Joyce Linnen, was aghast.

"Mamma," she said, "you got no shame."

"Diabetes got no shame either," Linnen responded.

While Linnen may be willing to stick her foot in her doctor's face, the region's racial history of segregation and oppression deters many African Americans from demanding better care or even asking questions. There is only one black physician in Georgetown County — a cardiologist — and minorities are generally underrepresented in the field of health care. White physicians, however, are quintessential authority figures, to whom deference is considered "proper behavior." Eliza Linen, a black nurse at Georgetown Memorial Hospital, says that some African American patients are so intimidated that they will ask questions only after the white doctor has left the room. Such passivity is bad for any disease, but particularly for diabetes, where the patient's knowledge is essential to treatment.

Moreover, black patients' faith in white physicians may not always be well placed, for racial slights can play out in ways both subtle and profound. Parents, for example, say that doctors who order insulin pumps for white children are reluctant to do so for their own, and the diagnosis of minority children is often delayed — a sign of ignorance or just lack of commitment. "Most of the kids here have to go into coma before [doctors] know they have" diabetes, says Pamela King, whose son, Kendrick, at the age of three, was initially diagnosed with an upper respiratory infection and was not admitted to the hospital until he had fallen into a coma, when he was finally diagnosed with type 1 diabetes. A doctor asked King if she had insurance for her son's burial. Kendrick is now twelve, but he has to travel to Charleston, sixty miles away, for care. "Doctors here do not understand juvenile diabetes," his mother says.

Racism could also play a role in amputations, according to Carolyn Jenkins, an associate professor at the Medical University of South Carolina's College of Nursing in Charleston and a principal investigator for the REACH program. Amputations are not unique to blacks in Georgetown. Each year, 82,000 diabetics in the United States lose a foot or leg. But Jenkins, who is white, believes that an African American in Georgetown who needs a toe or foot amputated may have his leg cut up to his knee. White surgeons think that such a patient cannot take care of himself and will eventually need additional surgery, so they amputate prematurely.

Compounding the problem, according to health officials, is that African Americans traditionally haven't had the resources for the kind of preventative care necessary in diabetes. Medicine, in this view, is not about promoting health but treating a serious illness or an acute disease — a mindset that assumes brutal logic in an impoverished community like Georgetown. About 17 percent of its 58,924 residents live below the poverty line, but that number increases to 31 percent — almost one in three — for African Americans, according to the U.S. census data for 2000. In 2003, the Saint James–Santee Family Health Clinic opened in a rural section of the county for uninsured or underinsured residents, providing free medical supplies for up to a year, conducting basic physical exams, and educating patients on caring for themselves. "We have a lot of diabetic patients," says Verna Grant, the clinic's nurse. "We ask who their doctor is, and they just say, 'We go to the emergency room.' Health care and medicine — they can't afford it."

Financial necessity also continues to require blacks to buy cheaper, less desirable food, as a visit to the Wal-Mart in Georgetown makes clear. In the meat section, huge packages of pork tails, layered in grease, fat, and salt, cost $1.97 a pound, while high-fat ground beef runs $1.98 a pound. Cured fatbacks and pork hocks, stacked in meat coolers like cholesterol towers, are inexpensive as well. Boneless chicken breast, however, is $3.10 a pound; better ground sirloin costs $2.98 a pound. Similar patterns are found throughout the store. Canned foods in light syrup are more expensive than those in heavy syrup; low-sugar cookies and sugar-free ice cream cost more than their sweeter counterparts; low-carb wheat and fiber bread costs $2.42 for twenty slices — compared to white enriched bread at 78 cents for thirty slices. "It's hard to eat healthy on limited funds," Linnen points out.

Cost isn't the only reason for soul food's popularity. It also connects African Americans to part of their history, a tradition that honors resilience, creativity, and community; and for that reason it plays a central role at church socials, reunions, and other gatherings. "Food was the one thing we could control," says Jeannette Jordan, a black dietician from Charleston. "It's connected to survival."

Linnen knows that diabetes will ravage her community unless people change the way they eat, but that requires breaking powerful traditions. The phrase "high on the hog" referred to southern plantation owners, who ate the best part of the hog, while the slaves got the fattiest pieces, often used to spice up other foods. Ham hocks or pig tails seasoned beans; fatback meat was mixed with collard greens. Chitterlings, hog jowls, pigs feet, and neck bones, as well as cornbread, biscuits, and candied yams, all remain inexpensive, high-caloric staples.

High-fat foods could be absorbed by blacks in Georgetown who labored in the fields, but now African Americans are more likely to work in a service industry job — ranging from bank officer to custodian — that burns far fewer calories. Their traditional diet simply contributes to weight gain, hypertension, and high cholesterol. At the same time, many African American women prefer zaftig figures, according to Jordan. "The average African American woman doesn't want to wear a size five or six, and three is out of the question," she says. "Even when they're at their correct weight, they want to gain weight to have bigger hips or larger breasts. Men want something they can hold on to."

Georgetown County's thirty-two African American churches have long been havens from deprivation and racism and a source of strength and solidarity. But the very power of faith — the belief that God is an omnipotent healing force — can undermine medical care. Some parishioners believe that they cannot trust both God and science, that following a doctor's orders would contradict their adherence to the Lord's will. According to health officials, some new diabetics resist treatment because they will not "claim" their disease: they are Christian, God is their healer, and their ability to function with the disorder, not just for weeks or months but years, confirms that Jesus is protecting them.

Florene Linnen tries to tell them otherwise. On a sunny winter morning, she drives to Nazareth AME Church in the county's Choppee section, across from a cemetery and surrounded by empty corn and tobacco fields. Inside the one-story building are stained-glass windows that depict Jesus hanging on a cross or genuflecting beneath a ray of sunshine. Ushers with white gloves escort parishioners to their seats. Members of the choir, dressed in white shirts, walk to the platform in the front. In a raucous celebration, the organist and drummer begin to play, the congregants clap their hands and bang their tambourines, and the choir sings "Everybody Needs to Know Who Jesus Is!" An old man in a dark suit and tie dances in the aisle, and a woman in a bright blue dress and gold hat takes the pulpit. "Just keep the families together, God!" she bellows. "Somebody's heart might be hurting, but you're a heart fixer! You're a mind regulator! You can do *all* things but fail!"

Linnen sits on the side, rocking a young girl in her arms. Then she is called to the pulpit, and standing behind the lectern she begins to sing, "If you're healthy and you know it, clap your hands." *Clap, clap.*

When she completes the song, she talks about diabetes. "It's a serious disease, but the good thing is that you can control it," she says. "In Georgetown County, we have the highest rate of amputation below the knees."

Linnen speaks without notes and makes eye contact with the congregants. She announces that she has "report cards" that diabetics should give doctors, specifying which medical tests they should get. "You are in charge when you walk into the doctor's office. You don't let him tell you. You tell him. You know your body. Don't let him say, 'You got a little sugar, you're a little high.' Don't let him do that." Too often, she says, diabetics will drink a lot of water before their appointment so they don't have sugar in their urine, but the A1c test will reveal your average blood sugar for the past three months. "So we cannot cheat with this. We have to be on the ball."

She recounts the story of a woman whose feet were so numb from diabetic neuropathy that she had lost all feeling in them. "She was walking on the hot pavement and didn't realize what was going on," Linnen says. "She smelled the smoke and looked down and both of her feet were on fire. She had to have them amputated." She told the women in the pews to buy comfortable shoes, "not shoes with that pointed toe, because if you look at our foot, it lay flat. It's not made for us to have high heels and pointed toes."

Following Linnen is Eliza Linen, the nurse from Georgetown Memorial Hospital, who lays out the consequences of poor control. "I'm seeing more and more young people losing their kidneys," she says. "Thirty-two years old. Thirty-five. Forty-two. Even nineteen years old. They're on dialysis. But it really starts here. It starts with yourself. If you don't start caring about yourself and what you put in your mouth, you're going to end up on dialysis and eventually across the street [in the cemetery]. We're losing patients every day on dialysis, and we're gaining them too. One dies, and one goes on."

The Reverend Michael J. Frost continues this theme as he recounts a personal illness, but his sermon has a very different twist. He is a trim, dapper man with short gray hair, the picture of health. But two years ago, he says, he was losing weight and suffering from a number of maladies, taking steroids for a muscle disorder and other medication for his heart, blood pressure, and diabetes. Nonetheless, his condition worsened, and no one knew why.

"But God, I believe through prayer, delivered me from those things," he says calmly. "I stand before you, a man who's been free of medication for two years. It's phenomenal to me. I look at some of your faces and you wonder if it's real or if I'm telling the whole truth, but I'm telling you that God is real, and I stand before you not taking a pill."

Frost raises his voice and talks faster, punctuating key words with barely constrained shouts. "I believe all of this happened because of prayer, and because getting before God and asking God and seeking God's faith and praying and admiring him for *who* he is and letting *Him* know that He is *ruler* over my life, letting Him know that He is *king* of my life, letting Him know that I serve a *powerful* God, that all *He* has to do is *speak* the word, and my soul shall be healed!"

Some congregants rise, cheer, and pump their fists. "*Yeah, yeah!*"

Frost takes a breath and reloads. "And I believe that even though I went to four doctors, and I had doctors corresponding back and forth, I believe that God was going to heal me one day. I believe that even though my body was truly racked with pain, you could touch me here and it would hurt. When I lay down at night, it would hurt. I would get up in the middle of the night because I couldn't sleep because when I lay down on the pillow . . . Pain after pain after pain. In the midst of all that I *still* didn't complain, but I *cried* out to the Lord! —"

"*Amen!*"

"To intercede in my life! —"

"*Amen!*"

"And over a period of time, God moved, and that same God that did it for me can do it for you. That same God can take away those pains and wipe away those tears. But you have to trust him. You have to learn to walk by faith and not by sight. Let us begin to walk this journey."

"*Amen!*"

It was a spellbinding performance. While Florene Linnen and Eliza Linen spoke convincingly about seeking medical help, they were no match for Frost's eloquence and fury, investing God with far more power than any endocrinologist or dietician. While Frost had earlier encouraged congregants to attend a health seminar, his description of the Almighty's healing power undermined any message of painstaking self-discipline or professional care.

After the service, he says there should be no contradiction between religion and science, for the Bible instructs that faith without work is death. "God created man and God created medicine," he says. "We pray first, and then we go see the doctor."

Ideally, that would always happen. But in reality, it doesn't, at least not for everyone, and it was easy to see why some pious diabetics don't "claim" their disease. If God can impregnate the Virgin Mary and raise Jesus from the dead, why can't He also restore beta cell function?

After the service, Florene and Joyce Linnen drive along dirt roads, passing scattered houses that have either been boarded up or need a roof, fresh paint, or new screens. They pull onto a dusty horseshoe road dotted with corroding mobile homes and lined with weeds and brush. The street is named Cemetery Court, though no tombstones are in sight.

The women stop the car and enter one of the dilapidated homes. In the back room, lying next to a bare light bulb, a color television, a tub of butter, an open bottle of ketchup, and a jug of orange juice, is

Bertha Rice. She is a sixty-three-year-old former schoolteacher, obese and immobile, with puffy bloodshot eyes. Confirming the high genetic risk of type 2 diabetes, she is part of a family that was destroyed by the disease. Out of fourteen children, thirteen were diagnosed with the disorder; three have already died of complications, and two are currently on dialysis. Her mother, an uncle, and two aunts also had it, with her mother dying on a dialysis machine. Her husband has diabetes; in 1996, he got a splinter in his finger, but by the time he went to the doctor, the finger had to be amputated. Since then, two toes have been cut off. At least three nieces have been diagnosed. Rice adopted a son whose biological mother had diabetes. The boy got it as well.

Rice has been taking insulin since 1985. Though forced to retire from teaching in 1994 for health reasons, she retained her insurance, so she can buy insulin for $40 a bottle. That keeps her alive, but not much more, because she also suffers from kidney disease, high blood pressure, retinopathy, and neuropathy. One day, while she was sitting in her wheelchair, a stiff wind slammed the door against her foot. She rolled to her bed and climbed in. A sister soon came over and noticed a bright red splotch at the bottom of the blanket. Rice's foot, badly cut, had bled through her sock, the sheet, and the blanket. "They had to cut off three toes, but it didn't bother me," she says.

She sees a doctor in town but has never heard of an A1c test. "The nurse says I'm high most of the time, but I don't know," she says. Asked why diabetes has had such an impact on her family, she shrugs. "When we were growing up, they just said, 'You had a little bit of sugar.'"

A commercial appears on the television, and all eyes turn. A woman holds up a FreeStyle Flash glucose meter and begins talking about its many attributes for diabetics. She is chipper and blond and beautiful, and the sleek silver device, which fits in your palm, makes life easy with a backlit panel display and illuminated test strip port for testing at night. Requiring the world's smallest sample of blood, it's "virtually pain free." The meter plus a year's supply of strips would cost $1,535, more than the value of Rice's house.*

---

* The price assumes that she would test four times a day, which is the minimum number to maintain glycemic control.

The commercial ends, the fantasy is over, and the stench of ketchup and juice and death hangs in the air. "Just a little bit of sugar," Rice repeats. "That's all they said."

Florene Linnen believes her work — and the entire REACH 2010 program — has made a difference, notably in amputations. In 1999, the number of amputations in African American males in Georgetown and Charleston counties was 79.1 per 1,000. In 2002, the number fell to 31.7, a decrease of 60 percent. Gains have also been made in blood pressure control and cholesterol levels. Linnen herself sees the progress in patients who now carry their diabetes report cards to their doctors or who buy skinless chicken breasts at the supermarket; or in amputees who, understanding the importance of exercise, use a cane to walk along a dirt road.

Linnen's own commitment to tight control has paid off in unexpected ways. In 2004, she was diagnosed with breast cancer and was treated with radiation, and she believes her rigid diabetes management helped her get through it. At sixty-two, she wouldn't mind retiring, but she needs to keep her job for the health insurance, and she wants to see the REACH program expand.

Unfortunately, neither may happen. Starting in 2005, funding for the program is being redirected to the "evaluation and dissemination" of its results. The five community health adviser positions, including Linnen's, will be eliminated unless another source of money can be found. Linnen says she will have to find another job until she is eligible for Medicare in three years.

"I guess I'm crazy because I'm not worried about it," she says. "I always believe there is another way."

CHAPTER 5

Rewarding Failure, Punishing Excellence

STEVEN P. PARKER sees himself as a medical detective, tracking clues to unlock scientific mysteries and unravel the human body's inner workings. As an internist, he must diagnose and treat adults with medication. It is the antithesis of surgery, which uses bold, decisive strokes and lends itself to a kind of therapeutic clarity. Parker's cases often involve chronic or incurable diseases, negating any resolution or closure. But the payoffs can still be significant. His greatest reward comes from making a difficult diagnosis and developing a superior treatment. "I apply scientific principles," he said, "while appreciating that much of what I do is also an art."

In 1998, Parker pursued what he thought would be the medical version of the American dream. After practicing for fourteen years and with the assistance of his wife, Sunny, he opened an office in Pensacola, Florida, confident he could be his own boss while doing what he most loved — helping patients one on one. Pensacola is not wealthy, but it's the largest metropolitan area in the Florida Panhandle and, with a sizable elderly population, it offered Parker many prospective clients. He found a good location, on the ground floor of a popular shopping complex near the offices of a family practitioner, a gynecologist, and a neurologist. His office had three exam rooms covering 2,000 square feet, a saltwater aquarium in the reception room, fresh coffee and snacks for visitors, and a small apartment upstairs for them and their two young children. Patients could drive right up to the front door.

The practice boomed. In three years, Parker had 3,000 patients, and he was especially popular among diabetics — who represented more than 20 percent of the clients — because he spent time assisting them with little things, such as the importance of clipping toenails to prevent infection, and making himself available when blood sugars went awry: his home number was listed in the telephone book. "Patients would call all the time," Sunny said. "It was our life."

What Steven Parker didn't anticipate was that the very attribute that made him a good doctor — his commitment to his patients — would also put his practice in a tailspin, place his own family at risk, and confirm that delivering good care to diabetics is an economic sinkhole.

Diabetes throws into sharp relief the perverse financial incentives that guide America's health care system. The Diabetes Control and Complications Trial proved that patients do far better when vigorously treated by physicians as well as nurse educators, nutritionists, therapists, ophthalmologists, and social workers. The team approach wasn't novel — Elliott Joslin had used it decades earlier — and the results weren't surprising. The chronic nature of diabetes, and its complexity, means it cannot be treated in the mechanistic fashion — diagnosis, therapy, resolution — used for other maladies. Diabetes is a constantly moving target, requiring time, energy, and money, producing endless streams of data and imposing the onus of care on the patients. Their greatest enemy is not high blood sugar, complications, or hypoglycemia. It is temptation, the lure of noncompliance, failing to take medicine, eat properly, or exercise regularly. A diabetic's treatment is equal parts applied physiology, information management, and behavioral medicine. There are no therapeutic shortcuts.

The country's health care system, however, is designed to treat episodic illnesses or acute disorders, the legacy of a bygone era. Infectious or pestilential diseases have long menaced humanity, at times with unimaginable force. In the 1300s, the Black Plague eliminated more than one-quarter of Europe's population. In 1918, the Great Influenza pandemic killed anywhere between 50 million and 100 million people worldwide. But by the middle of the twentieth century, the development of antibiotics had brought bacterial infections under control, lengthening lifespans and giving rise to a very different problem: chronic disease became the primary cause of mortality and morbidity.*

But reimbursements for medical care still favored episodic problems, with generous payments for emergencies, hospital stays, or one-time events and relatively little for prevention or disease management. Historically, physicians were paid a fee for service, charging what a patient could afford and ensuring that the wealthiest received the best care. Early in the twentieth century, reformers tried to institute national health insurance to protect Americans against sickness while promoting fixed payments to doctors and incentives for prevention. The effort failed, in part because doctors feared that government involvement would reduce their income. What emerged was private health insurance offered through employers, a benefit that expanded and was fortified during World War II, when wage and price controls led companies to offer health insurance to attract workers.

The insurance industry was first dominated by Blue Cross and Blue Shield, which were controlled by hospitals and physicians. Instead of a system to relieve the economic stress of workers, it was concerned with "improving the access of middle-class patients to hospitals and of hospitals to middle-class patients," according to Paul Starr's book on the history of medicine. Previously, doctors and hospitals had to wait months if not years for payment, but now reimbursements were prompt.

The system, however, did not benefit chronically ill patients, viewed as "incurables," who "tended to pile up in county, municipal, and state institutions." They were expensive to treat, their cases were depressing, and they failed to satisfy their providers' desire for closure. They were also seen as responsible for their own condition, for chronic diseases were often associated with poverty, criminality, and other social problems. Insurers rejected them for economic reasons, preferring to cover events beyond the control of the insured. Ma-

---

.* The CDC estimates that 90 million Americans have a chronic illness, though in 1996 the Robert Wood Johnson Foundation estimated that 112 million Americans had such a disease.

jor surgery or hospitalization fit that description. What carriers did not cover were non-emergencies, considered "controllable" events, such as office visits, X-rays, lab tests, or other forms of prevention. Paying for them would increase their use, creating what insurers call a "moral hazard" but is more easily understood as reduced profits.

As medical costs rose, insurers did indeed spend more money, but it was directed toward their base — hospital-oriented specialists, like surgeons, over those who worked in offices, including general practitioners and internists. This system was essentially replicated by the enactment of Medicare in 1965. By the 1970s, the federal government, seeking to control health care costs, asked the American Medical Association (AMA) to develop reimbursement codes for Medicare. The AMA recommended dollar amounts for procedures, which were adopted by Medicare and eventually by private insurers.

Reimbursement was based on the perceived difficulty of the work, the cost of tools or equipment, and liability expense. At the extremes, a simple procedure is paid little (trimming toenails: $10.15), while a highly complicated one is paid much more (the surgical reconstruction for a baby born without a diaphragm: $5,366.98; fitting a prosthetic, including consultations and pain medication: $12,923). Some surgeons, if they're in demand, will refuse insurance payments and charge market prices — $12,000, for example, for a gastric fundoplication, an operation that stops severe reflux of stomach acid.

Compensated at much lower levels were educational services, checkups, and office visits — the foundation of self-management and prevention. With help from the AMA, Medicare developed payment codes for these services as well, with reimbursement based on the credentials of the provider, the complexity of the problem, and the cost of living in the community where the service was performed. Prices were relatively low if the provider did not use any procedures or make physical contact with the patient. Thus, nurses, nutritionists, and diabetes educators today receive as little as $15 for a session in diabetes care, while an endocrinologist may receive $100 for a diabetic's checkup. Though endocrinologists receive special training, their cognitive skills are not highly rewarded by the reimbursement guidelines — which explains America's shortage of the very specialists needed to treat the diabetes epidemic. A study in 2003 concluded that, with only 3,623 endocrinologists in the country, the average waiting time for an initial visit — thirty-seven days — was longer than for any other specialist.

Insurers say that the amorphous nature of preventative medicine and chronic care makes its benefits difficult to quantify. It's much easier to recognize the advantages of, say, an appendectomy than to measure the value of near-normal blood sugars; it's difficult to demonstrate the added clinical value of a $6,000 insulin pump compared to a regimen of multiple daily injections.

While measuring the value of prevention may be problematic, the costs of complications are well known, particularly to the federal government. C. Ronald Kahn, the president of the Joslin Diabetes Center, estimates that it costs $10,000 a year for a diabetic to receive good care and to stay under good control. If the patient develops nephropathy, Medicare pays about $35,000 a year for kidney dialysis, and some 45 percent of dialysis patients have diabetes. "Even if I can't prevent you from going on dialysis but just delay it, the system is saving $35,000 a year — not to mention that you're better off, you're more productive, and you're not going to a dialysis center three times a week," Kahn said. "The federal government should have a lot of incentives to prevent that outcome."

Employers and insurers are also reluctant to invest in prevention because workers who change jobs change carriers as well. The benefits of maintaining good health would then accrue to a competitor. But this view is also shortsighted: departing employees are replaced by new ones, so each company and its insurer risk inheriting the very problem — a patient with an undertreated chronic disease — they thought they could ignore.

Another trend that emerged in the 1980s — the managed care movement — was a further blow to people with chronic conditions. With an emphasis on controlling costs, managed care made efficiency a priority, forcing providers to maximize the "throughput" of patients. Health care became a volume business. In the late 1980s, most doctors believed that seeing thirty patients a day pushed their limits, but by the 1990s, that number was common at many health maintenance organizations. Inevitably, the doctors' time with patients diminished. In 1997, physicians spent on average eight minutes per visit talking to each patient, less than half as much as a decade earlier. Treating the chronically ill, including diabetics, was still good for business as long as they could be treated quickly. But providing good care was bad for business because those patients consumed too much time, cutting into total volume and reducing revenue. As Kenneth Ludmerer wrote in 1999, "Only a few years ago, and throughout the history of medicine, attracting sick patients needing time and expertise had been considered a compliment to the skills of a good doctor, not a penalty." Thomas Bodenheimer, writing in *JAMA*, noted that clinicians "routinely experience the tyranny of the urgent," in which acute illnesses take priority and chronic diseases are neglected. "Too often, caring for chronic illness features an uninformed passive patient interacting with an unprepared practice team, resulting in frustrating, inadequate encounters."

The time crunch has been particularly hard on diabetics. Most are treated by general practitioners, who spend about eight to twelve minutes a visit with them — not enough time to examine blood sugar records, adjust medication, and screen for early signs of retinopathy, neuropathy, or microalbumin, let alone discuss the disorder's complex emotional jungle or adopt the experimental use of new technologies such as continuous glucose monitors. Time, in fact, is one of the principal reasons that insulin is not prescribed often enough for type 2 patients; considerable effort is needed to help them overcome their reluctance to take injections and to teach them proper techniques. Time is needed for all patients when they call or e-mail blood sugar results; unlike lawyers, who can bill for such calls, doctors and nurses cannot. Maintaining glycemic control increases the risk of hypoglycemic episodes that require emergency care or hospitalization, which requires further time from the doctor.

In this environment, it is not surprising that the physicians and clinics that spend the most time and resources caring for diabetics suffer the worst financially. The Joslin Center, for example, should be a financial juggernaut. It has more patient visits — about 100,000 a year at its Boston headquarters and various satellites around the country — than any diabetes clinic in the world. Founded it 1898, it has the greatest "brand name" in diabetes. It has more board-certified physicians with expertise in the disease, and a larger staff of certified educators, than any other center. It employs nurses, dieticians, exercise physiologists, mental health experts, and even child care specialists, who play with youngsters while their parents meet with the pediatrician.

But third-party reimbursements do not fully cover these services, and Joslin loses money almost every time a patient walks through its doors. Fewer than 10 percent pay the "published rate." The rest are covered by insurers, which pay about 50 percent. The center estimates that it is reimbursed 70 cents for every dollar it spends. In 2004, the clinic suffered an operating loss of $8.1 million, which it covers mostly through private donations.* Ron Kahn said the clinic attracts specialists who spend more time with their patients than health insurance providers usually allow. "They don't pay attention to the limitations of the system," he said.

If the health care system doesn't reward good care, the inverse is also true: it creates incentives to provide bad care, because a neglected patient who develops complications becomes an economic windfall for providers. Diabetes is not unique in this regard. Relatively little money is spent on helping a pregnant woman stay healthy, for example, whereas huge sums are available for her premature baby. Nutritionists are poorly reimbursed by insurers who are spending heavily on the side effects of obesity. At a hospital in Palm Beach, Florida, state inspectors in 2002 found "massive post-operative infections" in its heart surgery unit, requiring patients to undergo more operations and longer hospital stays. Medicare's response: it paid the hospital even more money for providing the additional care — an extreme example of how poor hospitals receive higher payments than good ones.

These misaligned incentives are played out each day at the Hunterdon Medical Center in Hunterdon, New Jersey. Its Diabetes

---

* Elliott Joslin never had to worry about money. His mother, Sara Procter, was heir to a large fortune from her father's leather tanning trade. Joslin's inheritance left him "a millionaire several times over by today's standards," helping support his clinic, which almost certainly lost money.

Health Center, with 4,000 patient visits a year, specializes in education programs, which garnered the American Diabetes Association Certificate of Recognition. Its classes include individual and group sessions for carbohydrate counting, lifestyle issues, and insulin pump therapy; stress management and fitness programs; and even yoga for diabetics. Providing information is only part of its mandate. Educators "must help patients deal with everyday obstacles," said Carolyn Swithers, the center's director. "It's the emotional and psychological as much as the physiological."

Consider the nineteen-year-old college freshman whose diabetes was poorly controlled when she was brought to the center by her mother. When Mary Whitlock, a certified diabetes educator, asked about her control, she said, "I don't care."

Whitlock knew that neither threatening nor cajoling would work. "Forget about diabetes," she said. "What's the most important thing to you right now?"

The woman said it was dating and trying to meet the right guy.

Whitlock connected that issue with her medical needs. "If you're at a party and you're low, some guy could take advantage of you," she said. "If you're dating a guy and you're 450" — her blood sugar level — "you won't feel good and that could hurt your chances."

The woman confided that her eyes were red all the time. "People think I smoke pot," she said. Whitlock suggested they might be red because her blood sugar was always high. The woman said she would try to take better care of herself.

Such counseling, while time-consuming, can be effective. The center tracks its patients and says that after one year, educational sessions reduce average A1c's by 17 percent, to 7.08. "If you live with a chronic disease," Swithers said, "you need a cheerleader and a support person. The commitment of the educators is what makes the biggest difference."

But third-party payers are much less committed. Medicare pays the center $28 for a thirty-minute session with a nutritionist or nurse, both of whom are paid $35 per hour; when building expenses and educational materials are included, the sessions lose money. Overall, private insurers reimburse the center 80 percent of its actual costs; the center loses about $200,000 a year.

"Education has never been a big-ticket item," Swithers said. "We've never invested in prevention." Despite the losses, the Hunterdon Medical Center continues to support the clinic because, according to Swithers, it sees the benefit of its work. "We monitor everything we do, and we keep showing outcomes," she said.

While the diabetes unit operates in the red, it helps its parent organization make money by referring its patients to the center's more lucrative operations, such as diagnostic imaging, podiatry, and cardiac services. Heart disease in particular is prevalent among diabetics. Someone with type 2 has the same risk of a heart attack as a nondiabetic who's already had an attack, while virtually all patients with type 1 have significant arteriosclerosis by the time they're forty. At the same time, cardiac services are unusually profitable for hospitals, contributing up to 40 percent of their profits. This not only reflects the high volume of patients but also the hefty reimbursements for cardiologists, sophisticated diagnostic equipment, and lengthy hospital stays. According to one study, the average cost of bypass surgery in the United States is $20,673.

At the Hunterdon Medical Center, present contract rates with insurers determine reimbursements, so a patient who is admitted to the center is charged $1,085 a day; admission to the intensive care unit is $2,323 a day. Pre-admission tests for heart services, including X-rays and EKGs, are $463. A patient who enters the operating room, regardless of the procedure, is billed $828 an hour; the recovery room costs $364 an hour.

Poorly controlled diabetics generate an unusual number of referrals to well-paid specialists. These include vascular, cardiothoracic, and transplant surgeons, ophthalmologists, urologists, orthopedists, nephrologists, and neurologists. The irony couldn't be more obvious, or painful, to Carolyn Swithers. The health care system will not pay her clinic relatively small sums for prevention, which could delay if not avert expensive complications. Only after the diabetic body becomes damaged do payers recognize the seriousness of the disease and give physicians an economic incentive treat it. "I don't get it," she said. "I really don't get it. We are rewarding poor care."

The rewards for poor care, and penalties for quality service, soon became clear to Steven Parker. When he opened his office in Pensacola, he hired a part-time physician and three support staff; Sunny, his wife, was the office manager and bookkeeper. Four days a week they saw patients; most of them came by word of mouth and appreciated Parker's personal attention. One day when he was running behind, a nurse entered an examination room to find that he was helping an old man get dressed — buttoning his shirt, putting on his socks and shoes — even though the patient's caregiver was waiting outside.

On average, Parker saw eighteen patients a day in his office as well as several in the hospital, about a twenty-minute drive away. Some days, he'd have to cancel up to eight office appointments for a hospital emergency. New patients were scheduled for thirty- to forty-five-minute visits; follow-ups took fifteen minutes, sometimes thirty. Following Medicare's reimbursement code, he would grade these visits from one to five, based on the complexity and seriousness of the illness. In 1999, Medicare paid him $38 for a level three visit, $45 for level four, and $55 for level five; if he did not code correctly, Medicare could fine him. (The rate was at least 30 percent more in South Florida, which has a higher cost of living.) Thus, it was far more profitable to see four level-three patients in an hour ($152) than one complex level-five patient, who would require more time.

His type 2 diabetics, coded at level three, created the greatest financial pressures. To make a profit, Parker could allot them only fifteen minutes, though they usually took longer. Their average age was fifty-five, with some as old as eighty-five, and they typically had multiple ailments, such as hypertension, heart disease, hypercholesterolemia, hypertriglyceridemia, transient ischemic attacks, leg or hip arthritis, or back pain. Most were overweight and sedentary. In any given visit, Parker would have to address three or four medical issues, so he always needed extra time. Because diabetics take more medication than most patients, checking in took longer. Diabetics also require more prescriptions to be written, more drugs refilled, and more referrals to specialists. If prescriptions were renewed too quickly, insurers would reject them, leading to skirmishes between the insurers and the office staff. Some patients ordered supplies through the mail, but the companies, leery of fraud, demanded even more paperwork from the staff. "Their intentions were good, but they didn't understand how much work was involved," Sunny said.

What frustrated Steven Parker was that many of the problems for type 2s could be prevented or cured with weight loss, dieting, or physical activity, but making those changes required — from a provider's point of view — time and support, and he had neither to give. He tried to juggle his workload to make it more profitable. Patients, for example, could come in for four separate fifteen-minute office visits at $38 apiece, but they resented four separate visits. So the appointments would often run twenty-five or even forty-five minutes, which was doubly bad: Parker was not being paid for the extra time, and he was angering his waiting patients.

He could sometimes increase his reimbursement to level four, but he had to be careful. According to Parker, Medicare often threatened to prosecute physicians who billed for complex visits while not meeting the precise criteria for the higher code. The coding itself is so abstruse that consultants will teach physicians — for a fee — how to label services that maximize revenue while avoiding prosecution. In fact, threatened fines are so intimidating that some physicians "under code," which also violates Medicare regulations.

While most businesses can raise prices — and make them stick — if their product or service is in demand, Parker couldn't raise his prices. Half of his revenue came from Medicare, 30 percent from managed care, and 20 percent from traditional insurance or out-of-pocket payment, so most of his business was paid for by the government or insurers, which rarely negotiate. He could have rejected Medicare patients or opted out of insurance contracts, but unlike a specialist with a unique skill, he would have lost those patients to a doctor next door. He also noticed that other general practitioners and internists who were doing well ran patients through like cattle or made extensive use of physician assistants or nurse practitioners. "Even if I were the best doctor in the region, Medicare paid the worst doctor the same fee," Parker said.

His expenses, of course, were constant: salaries, rent, utilities, equipment and supplies, and liability insurance, which rose each year; but the "mysteries of Medicare," as Parker called them, created irregular payments and occasional cash flow problems. At times, the Parkers had to forgo their weekly paychecks, and they could not afford to get sick themselves. If Steven didn't go to work, the business didn't generate revenue. One time when he was sick, he continued to see patients while wearing a mask. He walked slowly around the office, often slumping, until Sunny forced him to see a doctor. He had developed pneumonia and ended up missing two and a half weeks of work. "It took us a month to catch up financially," Sunny said.

Money was so tight that the Parkers could not afford health insurance for the family. "We were terrified that there would be something catastrophic," Sunny said. "We always feared someone getting sick or breaking a bone and not being able to pay. The only thing we had that belonged to us was our home."

They concealed these financial problems from their patients, who assumed they had plenty of money. When one patient was told her co-payment was $7, she threw Sunny a $10 bill and said, "Here's your co-pay so you can buy a new car." In fact, Sunny drove a 1982 Oldsmobile; Steven, a 1989 Ford F-150 truck.

Parker had few ways to increase his revenue. He could see more patients or could fraudulently "up code," but neither of those was appealing. He was already working seventy to eighty hours a week, including hospital rounds and on-call duty. A lab would have brought in more revenue, but insurers already had contracts with other labs and wouldn't cover the tests. He saw that other internists and family practitioners were shutting their offices and taking jobs at large clinics. That wasn't attractive to Parker either: those doctors could not spend thirty to forty-five minutes with a patient but needed to see six an hour.

In his third year, with Steven's business expenses rising, the Parkers had to find new ways to save money. They eliminated their janitorial service, and Sunny spent time on the weekends washing the windows and vacuuming. Steven did his own transcription. "You would think that a physician could afford a typist, but no," he said. At home, they used firewood to heat the house instead of electricity.

That year, the business generated about $300,000 in revenue. After expenses, Steven made $45,000; he paid Sunny $12,000. With two young children, they realized the sacrifice was too much. "We were spending every hour we had at the office," Sunny said. "We came to the realization, what kind of future do we have to offer these kids? There was no hope." She recalls the moment they decided to shut the office. "We just looked at each other and said we don't have a choice. To see my husband give up his love, it was a huge, huge hurt." According to Steven: "I finally said, 'Enough pain.' It was my greatest disappointment."

His patients were disappointed as well. One man came to their house and thanked him for saving his life while he was in the hospital. Others gave him food, wrote him poems, or made him a plaque. One woman painted a mailbox with a message at the bottom: "We will miss you."

In retrospect, Steven wishes he had pursued a specialty in cardiology, pulmonology, or gastroenterology, whose procedures would have provided him with greater income. He now realizes he was naïve in his belief that all physicians do pretty well financially.

While he doesn't blame any group of patients for his own setback, he believes the health care system deters the treatment of diabetics. As an internist, he knows more about diabetes than a general practitioner, but he isn't a diabetologist, and it's impossible for him to stay abreast of all the new medications and technologies in the field. While the goals for a patient have been well publicized — as Parker dryly notes, "Even the trial lawyers know them" — few practical guidelines exist on how to achieve those outcomes. Only clinicians who devote all their time to diabetes are truly qualified to treat it. Otherwise, the system must be changed so that nonspecialists have an economic incentive to provide quality care. "To a great extent, the patient must manage it himself," Parker said. "I don't see the current health care system allowing more physician involvement."

Legal liability also deters doctors from treating diabetics. According to Parker, if a patient dies prematurely or suffers a serious complication, "the long-term attending physician better have a documented exculpatory explanation as to why the goal was missed. The average primary care physician is just a sitting duck for this kind of litigation. Why accept that risk?"

After closing the office, the Parkers moved to Mesa, Arizona, where Steven got a job as a "hospitalist," a contractor who works exclusively at a hospital and whose average annual salary is $160,000. He no longer has to pay rent, utilities, or salaries. He still works brutal hours — seventy-six a week — but he now gets four weeks of paid vacation a year, and he has health insurance for his family as well as a 401(k) plan. The work is satisfying: he can observe the results of his treatment within hours or days. "I can see that I save lives," he said. The very disorders he once handled — "chronic, intractable, incurable and vague problems" — are deferred to an office doctor. There is much less paperwork and fewer insurance hassles. "I do almost pure medicine," he said. "I don't have to watch the clock continuously because the patients don't have appointments." As Sunny said, "Life is nice."

But the trajectory of Parker's career is a troubling microcosm of the health care system. He could not make a living keeping people out of the hospital, but he now makes a very good living once they're in the hospital — a poor outcome for the patient, who must be treated for a serious problem; for the system, which must pay for that treatment; and for Steven Parker himself, whose American dream is now only a memory.

CHAPTER 6

You Have to Be Brave, or Else It Hurts

AFTER THE WORST NIGHT of his life, Garrett sits on his bed in Children's Hospital, and I try to say something that resembles normalcy. Even at the age of three, he's a baseball fan, and his friends have schooled him on the immortal rivalries of Red Sox Nation. "Hey, buddy, I've good news for you," I tell him. "The Red Sox won last night."

He smiles. "And the Yankees lost."

At home, Sheryl tells Amanda that her little brother, like her father, now has diabetes. Amanda considers the ramifications. "We're going to need a lot of orange juice," she concludes.

Garrett's medical team enters the room to discuss his care. It's quite an array of talent, about a half-dozen men and women in white jackets, including two young pediatric endocrinologists, David Breault and Jamie Wood. Other parents have told me about the nightmarish experiences they've had with their doctors, but Breault and Wood are knowledgeable and comforting. "We're here," Breault says, "because we care about children."

I hold Garrett but realize I can no longer protect him. I know of children who were diagnosed as infants, but my main thought is just how small he is — not only physically but how small his world is. A three-year-old can't rationalize what's happening to him, can't find comfort in religion or philosophy or literature, can't read *When Bad Things Happen to Good People.* His universe has been violated, and he has no way to understand it.

I was a teenager when I was diagnosed, but my adjustment was relatively easy because my older brother was diabetic. I knew the routine, and I always had good care. I've never thought the disease was "unfair." It was simply a part of life that I stoically accepted. Now, for the first time, it's not only unfair, it's cruel. There is nothing redeeming about a condition that afflicts someone so young and helpless and that robs a child of his childhood.

We are not typical patients, given my own background, but I can't say all of my thoughts are rational. "My main concern," I tell the doctors, "is that he's going to die in his sleep." That rarely occurs — that a diabetic fails to wake up when experiencing hypoglycemia — but I know the times I've staggered out of bed in the middle of the night, sweating. On one occasion, when I didn't wake up, Sheryl had to call 911. I believe I would have eventually roused myself (in such circumstances, counterregulatory hormones should raise your blood sugar), but you never know. Now I fear, against all reason, that Garrett is going to die in his sleep.

In the course of the day, we meet with dieticians, nurses, and doctors, and we hear about insulins, syringes, glucagon, glucose meters, and carb counting. I can't imagine how a parent unfamiliar with diabetes can digest all the information. I've had it for almost three decades, and I'm overwhelmed.

Our nutritionist is young and cheerful, and she asks us what Garrett normally eats and then constructs a meal plan. Garrett has never been much of an eater, shunning vegetables and refusing such staples as chicken nuggets and hot dogs. His diet consists mostly of carbohydrates — waffles, cereal, fruit, yogurt. But not all carbs are created equal — some raise your blood sugar more rapidly. Unfortunately, these are the very foods that Garrett favors, but how can you tell a three-year-old his diet is loaded with high-glycemic carbs?

When I call my dad in St. Louis, I realize that for all the advances in diabetes care, some things haven't changed. I believe my brother has already told my dad the news, and he is as reassuring as possible. "Don't worry," he tells me. "They'll find a cure for this." My mind flashes back twenty-seven years, to the night I was diagnosed, and my mom and dad are standing next to my bed. "Don't worry," my dad tells me. "They'll find a cure for this."

Before lunch, we need to test Garrett and give him a shot. I tell him we have to do this so he won't have bellyaches or headaches, so he'll be strong and healthy. He doesn't mind the finger prick, but then I draw up the insulin and prepare to give the injection. He squirms, pushes back, and screams. "We'll do it quick, I promise," I tell him. I've given myself thousands of shots and know just how to minimize the discomfort by manipulating the angle, thrust, and speed of the needle. It's all about finesse, and every diabetic develops his own style. But I've never given anyone else a shot and have no idea what will work best for Garrett. All I know is, he's fighting back, hard, and Sheryl and a nurse have to help me hold him while I bunch up the flesh on his arm, aim the needle, and plunge it in. It's out in a second, but speed doesn't matter. Garrett curls up his fist, leans back, and punches me in a blind, furious rage. I don't blame him.

I later recall what Elliott Joslin said in 1923 — "The pain of insulin injection is slight. Otherwise children would not so readily give it to themselves" — and I wonder, what planet was this guy living on?

Researchers don't know why the immune system attacks the beta cells, though they know that children of diabetics are at higher risk for the disease. Joslin tried to discourage two patients from having children, presumably because of the likelihood of their becoming diabetic. (The clinic itself, however, was a pioneer in assisting pregnant women.) Other physicians, particularly before home glucose testing was possible, deterred diabetic women from starting families, and I'm sure many patients over the years have opted out of parenthood for fear of passing on the disease. It's a valid impulse.

Now I've passed it on, and as I go home midday to take a shower, I just wish there were some way I could tell Garrett how sorry I am, how I never meant for this to happen. I want to tell Sheryl how sorry I am, how sorry I am she ever married me. I get home and find that one of our neighbors has dropped off a box of sugar-free cookies. It's a nice gesture but more than I can handle — a reminder of the diet restrictions that will now inform my son's life, a symbol of his new outcast status. I have to be strong in front of Garrett and Sheryl, but now I'm alone. Desperate thoughts run through my mind. No child should have to live with this, and I conclude it was a mistake to ever have had children. It was a mistake to even get married. What was I thinking? How stupid could I be.

I clean up, eat something, send some e-mails. Time passes, but not the pain.

I return to the hospital. Before dinner, I take Garrett to a game room, where a precocious sixth-grader with a backpack is coloring. Her name is Chanel, and she tells us that she's visiting her brother. "He has CF [cystic fibrosis]," she says. "He's seventeen, but developmentally he's only three. I read him books. He'd do the same for me."

She sees Garrett's wristband and asks him why he's in the hospital.

"Because."

"Because why?"

"Because I have diabetes."

The word stuns me. It never occurred to me to tell Garrett what he had, but Sheryl has. Now the word sounds so . . . adult, so ominous. Until now, I have never considered how its first syllable seems to foreshadow one's fate, hinting at a bleak and treacherous future. The word should not fall from the lips of a child so young.

We return to the room, where a minister has stopped by and left his card. I try to lighten my own mood, asking Garrett in jest, "Do you want spiritual guidance?"

"No!" he yells. "I want a snack!"

The doctors put Garrett on three shots a day. Fast-acting insulin before breakfast and dinner; slow-acting before bedtime. I realize that three shots are fairly standard for intensive therapy — I'm on four — but it still seems like a lot for a child. Each injection is going to be an ordeal, at least for the near future. (I remember the line from a diabetic comic: "I take six shots a day, four of which are tequila.") But I also know that even three shots may not be enough. If Garrett's blood sugar spikes after lunch, he may need an injection before that meal as well. We're also supposed to test him before each meal, before bedtime, and at 2 A.M. The prospect of round-the-clock monitoring, vigilance, and oversight is daunting.

We're to stay in the hospital only two nights. I assume that's quick because the staff doesn't have to teach us the basics, but that's also the realities of the health care system. A mother later tells me that Massachusetts General Hospital declined to admit her young son when he was diagnosed. The staff just gave her an overview, packed some supplies, handed her the help number, and sent them on their way.

Sheryl goes to the hospital pharmacy and buys Garrett's first load of supplies: syringes, two types of insulin, test strips, alcohol wipes. With insurance covering most of it, she pays $183.

As we prepare to leave in the early evening on our third day, a nurse hands me a piece of paper with Garrett's insulin doses. The past two nights, Garrett had taken Humalog before dinner and NPH before bed. Now, according to this sheet, he is to receive the opposite — NPH before dinner and Humalog before bed. I notice it only because I am familiar with the insulins, and I question the nurse. She walks out the door and returns ten minutes later.

"It was a mistake," she says. "It should be the other way around."

Taking fast-acting Humalog before bed, particularly in a small child who's hypersensitive to insulin, could have dire consequences. I'm more exasperated than angry. A pediatric endocrinologist later tells me that hospitals commit more errors on insulin than on any other medication. I suppose that's not surprising. Different insulins come in similar bottles, and the margin for error is small. I'm glad we're leaving.

On our first morning at home, I'm nervous as I draw up Garrett's insulin. The fluid looks so harmless, but I know its awesome power, even in fractional units — its ability to save, yes, but also to destroy. Severe hypoglycemia is dangerous for anyone, but particularly young children; bad lows can damage the development of their brain. A parent later tells me how unsettled she felt giving her first injections at home to her young son. At the hospital, one nurse would draw up the insulin; a second would check that the dose was accurate. The irony was bizarre: to ensure the safety in administering a potentially dangerous drug, medically trained professionals have safeguards and support. But parents fly solo.

I jab the needle into Garrett's arm. He says "Ow," but the first shot at home is uneventful.

In the morning, we're playing T-ball in the driveway when he asks me, "Daddy, when will my diabetes go away?"

I tell him I'm sorry, but it's not going to go away. My heart breaks a little bit more.

While children have never constituted a large percentage of diabetics, they have always given the disease its special poignancy, its distinctive sorrow. Elliott Joslin's own obsession with the illness is attributed partly to his experiences with youngsters — before the discovery of insulin — as they valiantly but futilely tried to survive. He was particularly affected by a fifteen-year-old named Alton, whom he tried to keep alive by diet. The boy, however, attended a county fair for two days and returned ill from overeating. Joslin went to his home, sat by his bed, and wouldn't leave, watching him die on the third day. Children never made up a significant part of Joslin's practice — less than 15 percent — but that was enough to leave an impression. The medal that Joslin gave to patients who had exceeded their life expectancy paid tribute to this group: a boy etched on one side, standing with his dog on a boat before a setting sun. Inscribed around the edge: EXPLORERS OF UNCHARTERED SEAS.

To the public, diabetic children may have evoked sympathy, but they were never a cause of national concern. Before the discovery of insulin, the disease was no different from other childhood illnesses with high death rates — typhoid fever, tuberculosis, pneumonia. After insulin, diabetics lived while other children perished (polio victims became the most conspicuous example), and the pool of diabetic children was too small to create any public urgency. In 1935 and 1936, the U.S. National Health Survey sampled 2.5 million people and reported a diabetes rate for children under fifteen of 0.38 per 1,000.

However, researchers now believe that while the diabetes epidemic is focused on type 2, a significant rise is also occurring in type 1, with the largest rate of increase occurring in children who are four or younger. They estimate that in Western countries, three to four children per 1,000 will require insulin by the age of twenty. The problem has received little attention in the United States, in part because the government has made no systematic effort, until recently, to count, estimate, or track the number of diabetics under eighteen. Only in 2000 did the CDC announce a study to count representative samples of young diabetics. Fortunately, other governments have made a greater effort. A registry of forty-four European countries, called the EURODIAB collaborative study, indicates that between 1989 and 1998, the annual rate of increase in the incidence of type 1 diabetes among young patients was 3.2 percent.

Anecdotal evidence in the United States confirms similar growth rates. The Yale Program for Children with Diabetes had 950 patients in 2004, an increase from 350 ten years earlier, and the patients were much younger as well. "When I started, the average ages were ten to twelve," says Jo Ann Ahern, the program coordinator, who's been caring for diabetic children for twenty-five years. "Now most kids are under ten, and ages three, four, and five are much more common."

While childhood obesity can be directly linked to the rise in type 2 diabetes, the factors driving type 1 remain unclear, though theories abound. A new environmental agent, such as a virus, or changes in childhood diets (such as increased exposure to cow's milk), have been indicated; but these possibilities remain unproven. Another theory, called "the hygiene hypothesis," suggests that improved sanitation has weakened our immune system, making us more vulnerable to autoimmune attacks. The immune system has two kinds of lymphocytes (white blood cells). The first responds primarily to bacteria and viruses; the second, to parasites. The two exist in balance. But what happens if a child isn't confronted with enough bacteria or viruses? In theory, this could cause the underdevelopment of the first type of lymphocytes and the overdevelopment of the second, causing the latter to "overreact," or become too aggressive, ultimately killing its own beta cells. While this remains speculative, some scientists are beginning to think that children should play in dirt.

What hasn't changed is parents' commitment to their children and their ingenious efforts, born of desperation, to provide care. Consider Larue Drager of Spokane Valley, Washington, whose daughter, Staisha, was diagnosed with diabetes at five. When her blood sugar plummeted, Staisha resisted food or drink. One time at midnight, she was hypoglycemic but again refused juice, glucose tablets, gel, or anything else. Despite the freezing cold, Drager sped to the corner store and told Staisha that she could have whatever she wanted. Staisha chose chocolate milk, and Drager thought the problem was solved — but her daughter still balked at drinking. Drager feared Staisha would soon be unconscious, so she asked if she wanted to go to an empty parking lot and spin "broadies" on the snow and ice. Staisha agreed but was still not entirely cooperative. At the lot, she drank only when her mother spun the car. The night wore on, with the car skidding crazily across the slick asphalt. *Spin, drink. Spin, drink.* The girl's blood sugar rose to a safe level, though Drager was not comforted. As she recalled, "I thought my life ahead was full of out-of-control broadies in the parking lot." She says she can now laugh about it and, more important, has found better ways to treat low blood sugar.

Parents will go to extraordinary lengths to minimize their children's pain. In one case, a father was so dismayed when his newly diagnosed two-year-old son resisted his shot — a nurse had to chase him around the room — the father called out, "I have diabetes too! We're going to take our shots together." And he did, though he didn't have the disease, injecting himself with a syringe full of water, which indeed calmed the boy. But when they went home, the father continued taking shots for three years! He finally discussed the matter with the boy's endocrinologist, who urged him to tell his son the truth. At a meeting that included the doctor and a psychiatrist, the father began talking about the injections when the boy interrupted: "Daddy, why do you take shots? You don't have diabetes."

As treatment has become more sophisticated, with multiple daily injections and blood tests, it has also become more emotionally and physically taxing, and parents recognize that the drudgery can be endured only with humor and grace. When Michelle Chase tried to check her fourteen-year-old son's blood sugar at 2 A.M., Ryan sat straight up and, while asleep, delivered a right hook across her face. He then lay right back down as if nothing had happened. Chase, livid, returned to her bedroom, woke her husband, and yelled, "Next time, you test his blood sugar at 2 A.M.!" Her husband suggested wearing a snowmobile helmet. The next morning, Ryan asked his mother where she got that mark on her face. In time, Chase could recount the incident as an amusing anecdote. "I always remember this instance," Chase said, "when I get angry about diabetes and remember that God is making sure to add a little bit of humor to our lives to even out the bad stuff."

In their own way, children also inspire their parents. The kids, to be sure, aren't happy about their condition, and they will often resist its treatment. But most adapt — some immediately, others later — and their resilience and good humor reveal new sides to their personality. Katie Edwards is a mother who was diagnosed when she was two. As a child, she attended the Clara Barton Camp for diabetic children in North Oxford, Massachusetts. As an adult, she took her two children to a camp alumni day to share one of her favorite places.*

Edwards's five-year-old daughter, Ariana, loved visiting the camp so much that she was disappointed she couldn't go. Several months later, Edwards began noticing the symptoms in Ariana but tried to ignore them. One evening, she took her kids to the Laundromat, where Ariana drank an orange soda and then began vomiting. Edwards rushed her home, put her in a bath, and tested her blood sugar. She broke down when the reading was displayed. As she later wrote: "I left the bathroom. I did not want her to see me upset, and I fell to my knees in the living room, wondering how I was going to do this alone, having left my first husband a year before. And then my daughter's sweet little voice, almost songlike, came from the bathroom. 'Mommy, does this mean I can go to your camp now?' I was immediately blown away by this; here I am afraid of changes, and she was just looking forward to all that was open to her."

Two weeks after Garrett's diagnosis, his silver bracelet arrives in the mail, which Sheryl had bought online. It has the red insignia for diabetes, with the back inscribed with Garrett's name, telephone number, and the words "Insulin Dependent." I never wore a medical alert bracelet or necklace, carrying a card in my wallet instead. But every diabetic child should, for obvious reasons. Nonetheless, I can't help but feel a deep sadness when I see Garrett's bracelet. He is now marked and identified. It's not that I'd been in denial, but somehow his condition assumes a new permanence.

We explain to Garrett why he needs to wear it. I half expect him to refuse or to tug at it like a wild animal trying to free his leg from a

---

* The camp itself, named after the founder of the American Red Cross, began in 1932 as a joint venture between Elliott Joslin and the Women's National Missionary Association of the Universalist Church. Many similar camps followed, but Clara Barton became known as the first "hospital in the woods" for diabetic children.

trap. But he complies. Then one day when we're sitting on the couch, he says something that shakes me.

"My bracelet makes me sad."

If my greatest fear is Garrett's dying in his sleep, my second is that diabetes will change his personality, that he will somehow internalize the cosmic unfairness of life and allow his condition to dampen his spirits, to silence his laughter.

"Why does it make you sad?" Sheryl asks.

"It's not cool," he says.

"We'll buy you a cooler one," Sheryl assures him.

He doesn't raise the issue again.

We explain his diagnosis to his preschool teachers, his friends, and our neighbors. The more people who know, the better; and we don't want Garrett to feel as if he has something to hide or to believe he should feel shame or embarrassment. Still, our policy has limits. In the winter, we want to enroll Amanda and Garrett in a one-day ski lesson on a nearby mountain. It would be their first time on skis, and we're excited. But the minimum age is four, making Garrett too young. Nonetheless, we think he's athletic enough to hold his own, so we fib about his age when filling out the online application. Then it asks about health issues. I'm prepared to type in that he has diabetes, but Sheryl stops me. We'll be right there, she says, and the ski school may prohibit diabetic kids from its program. Why take the chance? I concur and type in that he has no health issues, but I wonder about the precedent: let's see, we've lied about his age and we've lied about his health. Can we introduce any other deceptions?

Both Amanda and Garrett flourish in their classes, whipping down the bunny hill on a sun-splashed day, and we take them for a second lesson a few months later.

One of our first fears was that Garrett's preschool, Carter Center for Children, would no longer allow him to attend. Under the Americans with Disabilities Act, schools that receive public money are not only obliged to accept diabetics but must also provide basic health services, including testing a child's blood sugar. Carter, however, is a private school, so we are grateful when its directors, Barbara Carr and Peggy McDonald, assure us that Garrett is still welcome. In fact, they and their staff make extraordinary efforts to help us, writing down his snacks each day, touching his brow while he naps to check for hypoglycemia, and recognizing changes in his personality when he's low.

We don't want to ask the school to test him, so each day I drive to Carter before lunch and do it myself. I fear that he is going to loathe my visits because they will make him stand out. On the first day, I ask if he'd like to be tested in a different room, away from the other children. He says no. He actually wants to test in front of his friends, who, sitting in a circle, are instantly captivated by the process. Some of the kids want to test their own blood, and I'm reminded of the woman who takes her meter to a bar, where she and her nondiabetic friends test their blood sugars: whoever is lowest has to buy the next round of drinks. I tell Garrett's friends that the meter is only for him. On later visits, he selects which friends he wants to have watch him, and his "tester" — as he calls it — becomes the hottest ticket in school. One time, after recording a blood sugar of 125, I say, "Awesome number! One twenty-five." The other children, picking up the cue, start dancing around and singing, "One twenty-five! One twenty-five!"

Garrett has always been stubbornly self-sufficient. If he sees Sheryl or me doing something around the house — making coffee, sweeping, vacuuming — he wants to do it by himself. While Amanda wants help picking out her clothes and getting dressed, Garrett rejects our assistance. His independence will serve him well, for the sooner he learns the mechanics of his care, the better. It will be at least four or five years before he can give himself a shot or — if he goes on a pump — before he can change his infusion set and administer his insulin. But within a few months, he wants to poke his finger himself. He picks up the lancet, jabs himself, and pushes out a drop of blood. We never teach him; he simply learns by watching, and it gives him some feeling of control. (One mother tells me that her son licks the blood off his finger after testing, which may save on tissues but is not a practice I encourage for Garrett.) Several months later, Garrett puts the strip in the meter, pokes his finger, and tests himself. It's like watching your child's first step, but it's even more dramatic. He is learning how to survive. He still doesn't understand what the numbers mean, but the episode practically brings us to tears.

98

Garrett surprises us in other ways. One day he points to a banana and asks, "How many carbs?" Sheryl and I look at each other as Garrett laughs because he knows he's impressed us. We try to explain that he takes insulin to keep his blood sugar in the right range, that food will raise his level, and that exercise will lower it; he quickly puts the pieces together. After eating, he'll remark, "My blood sugar is this high," raising his hand to his neck. He also figures out how to exploit the situation: when he wants a bigger snack, he screams, "I'm low! I really am!" A quick blood test either confirms his instinct or reveals his deceit. One evening, Garrett rides his bike while I try to keep pace on foot. When we get back to the house, we're both breathing heavily, and Garrett looks at me and asks, "Are we low?"

Having diabetes is having an endless conversation with your body. I've spent much of my life trying to figure out what my glucose level is, but now I need to channel Garrett's as well. Blood tests help, but they are simply a snapshot of a constantly moving target. And while the tests are not as onerous as shots, they're still an intrusion. The hospital instructed us to test Garrett each morning at 2 A.M., at least for the first several weeks. Some parents do this regularly. Our first night back, I wake up in the middle of the night, walk into Garrett's room, and prick his finger. The poor guy screams, shocked by the assault. I squeeze out the blood and do the test. The following night, I walk into his bedroom at the same hour, but this time his hands are curled up against his body. I leave his room and rarely test him again while he's sleeping. We check him before he goes to bed and make sure he eats a good snack, but I don't think a child should sleep in fear. He should dream about what he loves — Power Rangers and bicycles — not finger pricks.

Sheryl and I react carefully to his numbers. We discreetly wince if he's over 300, for we don't want him to believe there are good numbers or bad. Every number is good because it gives us information to make the right decision, so even if he's over 300, I say, "Thank you, bud. Good job."

Amanda becomes keenly interested in his readings. She's always been protective, and now she anoints herself as the unofficial custodian of Garrett's blood sugar scores. "What was he?" she demands, and — without our teaching her — she somehow learns what his range should be. On one of the few nights that Sheryl and I go out, Sheryl's sister watches the kids and tests Garrett. The meter reads 61. Amanda springs into action. "He needs his snack now!" she instructs. Garrett eats his cookies, and a possible emergency is averted.

Before Garrett's diagnosis, I interviewed a woman whose son was diagnosed in the early 1960s. When it came time for his shot, the little boy would run around the house, and the woman, with a syringe in one hand and a cotton ball in the other, would chase after him, grab him, hold him, and with him kicking and screaming give the injection. Forty years later, at shot time, Garrett sometimes runs around the house, and I, with a syringe in one hand and alcohol swab in the other, chase after him, grab him, hold him, and with him kicking and screaming give the injection.

This is why I no longer celebrate the progress in diabetes care.

The week after Garrett's diagnosis, I happen to interview Jo Ann Ahern, the specialist in childhood diabetes, at Yale and ask her how many daily injections her patients take. She says two. I tell her Garrett's on three, which she says is unnecessary for most young children. When we see Garrett's pediatric endocrinologist for the first time the following week, he agrees to consolidate the two evening shots into one before dinner. As Ahern predicted, the change has no impact on his control. For most adults, the lifestyle difference between two shots and three is minimal, but for children it's huge. Part of me wonders why Children's Hospital didn't simply start Garrett on two shots. Perhaps its experience shows that three produce better results, but the Yale program has evidence that a less grueling schedule is just as good. Unfortunately, there is no best practices manual in diabetes. Much of it depends on your clinic or doctor. Money plays a role in good care, but luck is another factor. In our case, a lucky conversation with Ahern saves my son hundreds of unnecessary shots.

Which is not to say the hundreds he does take go well. They often don't. In those first few months and beyond, Garrett sometimes throws a pillow or a toy after his injection, or he hits me. (He seems to take pity on Sheryl, sparing her any violence.) He complains when it hurts and sometimes screams, "I don't want to have diabetes!" I tell him that he has to take shots so he can grow up to be big and strong — which is literally true. By the 1960s, doctors realized that many young patients experienced stunted growth, owing, no doubt, to poor control, and they were called "diabetic dwarfs." That means little to Garrett, whose only focus is the needle. "I don't want to be big and strong!" he shouts back.

We give Garrett a choice of where he wants to do his shot, and he develops a routine in which he extends both arms straight in the air, then makes the sound of an explosion, slowly dropping his fingers like fireworks falling from the sky. The fingers then land, as if guided by the wind, on the site of his injection, either a leg or an arm. It's an elaborate stalling tactic but entertaining nonetheless, all designed to give Garrett some stake in the process. As long as Sheryl and I are home, he demands that we sit on either side of him, then he says, "Switch," forcing us to change positions. We obey. The maneuver delays the inevitable while allowing him to have some control over the experience. Garrett holds Sheryl's hand while I give him the injection. If Sheryl is out, Amanda holds his hand.

Allowing Garrett to feel in charge of his life is a theme throughout this first year. When we serve him his usual breakfast — waffles — he now insists on more butter and syrup (we used low-cal even before his diagnosis). It doesn't matter how much he has already; he likes the power of demanding more, and he isn't going to eat unless we comply. If Amanda makes such a demand, we simply say, fine, you won't eat breakfast this morning. But that isn't an option with Garrett, and I suspect he understands his new leverage. We hate to give in but know we have to pick our fights carefully. So we skimp on the first round of butter and syrup, then give him more when he demands it.

I wonder how much his disease influences his behavior. When he misbehaves — whether it's acting out at school, not playing well with his friends, or being disrespectful — is it caused by his diabetes or is it normal for a young boy? When does the diabetes part of his personality end and the other part begin? I ruefully accept that there is no separation, no way of ascribing causality — that the boy and disease, for better or worse, are inextricably one.

Food is not so much a battleground as a difficult balancing act. On the one hand, we don't want Garrett to feel deprived; on the other, we can't allow him to eat at will. We explain that cookies, cupcakes, and other foods with sugar need to be avoided or eaten in small quantities. He's never had much of a sweet tooth, but making something forbidden increases its appeal. We compromise. On Halloween, we let him keep several chocolate bars, which he can eat for special snacks. At snack time, if his blood sugar is on the low side, we'll give him Oreo frozen yogurt. At birthday parties, he has a choice of ice cream or cake.

During the first spring and summer, an ice cream truck rambles through our neighborhood with distressing regularity, its cloying jingle announcing its arrival. Once the tune is heard, Garrett drops what he's doing, bolts out the door, and races after it with a dozen other kids. I don't think Garrett likes the treat so much as the excitement of the truck. By summer's end, I want to offer the driver $1.25 for the ice cream and $100 to stay out of our neighborhood.

Sometimes Garrett's friends will remind him of his limits. When a neighbor with a Popsicle sees him, he yells, "You can't have this because you have diabetes." Another friend at his school says the same thing when cupcakes are distributed. Garrett doesn't seem bothered, but Sheryl and I blanch. One night in the kitchen, Sheryl looks at me sadly and says, "They're going to tease him."

At other times, Garrett's maturity impresses us. One day, the director of his preschool tells us that the children were given Popsicles for a special snack. When they handed one to Garrett, he declined. "It will raise my blood sugar," he said. On other occasions, he tells his teachers that they have to call me to see if he can have some confection. But we hit a snag when the preschool begins serving marshmallows for a "campfire snack." Garrett has his own snacks from home, but he has the option of eating those from school, usually crackers, pretzels, or such. He eats the marshmallows, likes them, and for several days comes home with a high blood sugar.

Regardless of Garrett, I'm dismayed at the school's choice of snacks. I'm not a food fascist, but with childhood obesity skyrocketing, all schools should emphasize nutrition. I tell one of the directors that Garrett shouldn't eat the marshmallows anymore and suggest that they aren't great for the other kids either. She agrees. The marshmallows are nixed for everyone, and Garrett does not feel deprived. I also notice that some of Garrett's playmates begin bringing some of his foods in their own lunches: the exact same sugar-free juices, low-carb popcorn bags, and low-cal cookies. It's a beautiful sight, removing any stigma that Garrett might attach to those items.

But we can't stop the tide of sweets. Play dates, holiday gatherings, school, whatever — a culture of glucose prevails. The worst are birthday parties, when parents engage in the candy equivalent of an arms race, stockpiling chocolates, jawbreakers, and taffy and then strafing the kids with them on their way out the door. At one party, the kids break a piñata and then scramble after Tootsie Rolls and sweet tarts like they're gold bullion. We let Garrett have one piece, then we put away his bounty; thankfully for us, he forgets about it.

I also become aware of our town's lollipop conspiracy. Lollipops are given away at the post office, the bank, the barbershop, shoe stores, and street fairs. They're displayed prominently in convenience stores, drugstores, and supermarkets. I half expect the dentists to be giving them out. Garrett likes them, and while I try not to give in, it's hard to always say no. Do I allow him that treat so he feels like a normal kid, or do I deny him, causing him to feel isolated or angry? I have a better idea: I want the town to pass a lollipop moratorium or at least create a few lollipop-free zones.

Like any growing boy, Garrett's often hungry, so we develop a game. If he's already had his snack but wants more, he has to run wind sprints — he calls them "wing sprints" — around the house. We also do this on vacation in a hotel suite, and Garrett actually enjoys it — dashing from the living room, through the small kitchen, to the bedroom, and back. The other guests must think we're training for the Olympics. But sometimes we still have to say no. We have Nutri-Grain cereal bars, which are loaded with 28 grams of carbohydrate. Garrett can have one when his blood sugar is low, but now he wants another. When told he can't, he stalks off in a huff. "If my little boy has diabetes," he says, "I'm going to let him eat a Nutri-Grain bar."

His endocrinologist at Children's Hospital comes highly recommended; at our first appointment, he is reassuring, avuncular, and willing to do whatever he can to help Garrett. He talks about setting realistic goals and adjusting them over time. Garrett's glucose target, for example, should be between 100 and 200, which is higher than what other pediatric endocrinologists recommend but I believe makes sense. However, the doctor says one thing that makes me cringe. In gathering family history, he asks about Garrett's siblings.

I say he has an older sister.

"Is she healthy?" he asks.

Healthy? The implication is that Garrett is unhealthy, as am I. This is more than semantic quibbling. How diabetics are viewed — as fundamentally healthy with an inconvenient disorder or as chronically unhealthy with a progressive, debilitating disease — influences everything from health care reimbursements to medical research to public policy. More important, how patients view themselves can shape their entire outlook on life. I've never considered myself unhealthy, and I certainly would never want Garrett to feel that way. It undermines a psychological imperative: the need to feel confident, self-reliant, optimistic, and — yes — healthy. I am surprised that a pediatrician, clearly sensitive to other matters, uses such a word.

"Yes," I say. "Our daughter is healthy."

During that first year, I often tell Garrett how big and strong he's getting and how well he kicks the soccer ball, hits the baseball, or throws the football. All of which is true but beside the point. I never want him to think his diagnosis has diminished him physically. He is too young to appreciate that diabetics have climbed Mount Everest, swum in the Olympics, or even fought in Iraq. What he knows is his own world, and I want him to know that in that world he is flourishing.

And at night, after we read books or watch television, I tell him something else: "I'm very proud of you, Garrett. Do you know why?"

I answer my own question: "Because you're very brave."

I tell him this for several nights. Finally, after I say "because you're very brave," he has a rejoinder.

"I'm brave for tester," he says, "but I'm not brave for shots." I hug him a little tighter. "Yes, buddy, you're very brave for shots."

A few weeks later, when I test Garrett at preschool, a girl who is watching says, "I don't like shots."

"You have to be brave," Garrett explains to her, "or else it hurts."

A mother tells me that after her twelve-month-old son was diagnosed, she and her husband didn't go out at night alone for five years because they didn't trust a babysitter. They still had evenings out, but they had them alone or with separate friends, and their spouse stayed home with the child. Sheryl and I are determined not to be held hostage. After a few months, we decide to go out and ask our babysitter to watch the kids. We don't ask her to test Garrett but leave her with instructions on warning signs for hypoglycemia, snack times, cell phone numbers, and emergency listings. We leave, eat dinner, come home. Garrett is fine, of course. We test him and put him to bed. We're pleased for the few hours alone but also know that the old days — when we could take a train to New York for the weekend and leave the kids with friends or family — are over.

Several months after his diagnosis, Garrett tells a neighbor about the hospital, mentioning the adjustable bed, the television, and the playroom. I take this as a sign that he has no bad memories of the experience. But a few days later he tells someone else that the hospital hurt his left arm, recalling — though not knowing the words — the wooden block that immobilized it for the IV draw. A full year later, he still describes how the hospital hurt his arm.

New diabetics, in a period known as the "honeymoon phase," still have residual beta cell function; injected insulin is more of a supplement than a replacement. This makes control easier but also places patients at greater risk for hypoglycemia, particularly young children, who are extremely sensitive to insulin, don't recognize lows, and don't have conventional symptoms. Garrett, for example, is asymptomatic even when his blood sugar is in the 50s, which is quite low.

For the most part, we avoid any problems until one night in December. I suspect something is amiss when his blood sugar is uncharacteristically low before lunch. He's fine at dinner; then he usually plays and insists on staying up late. But now he's lethargic and just wants to sleep. We check him and he's 47, the lowest he's ever tested. We get him a juice box and urge him — loudly — to drink. (I've heard doctors all but yell at patients to keep them conscious.) Garrett sips about half the box but is too tired to finish. We wait and test him again; he's still in the 40s. We try to get him to finish the juice but he refuses. So we give him milk. He takes a few gulps but just wants to sleep. We try cookies, crackers, ice cream. We go back to juice. Anything. He eats this, drinks that. Given how small he is, he should have the necessary calories by now. After about an hour we test again — and he's 50. What's going on? We get out the glucose gel, a sugary toothpaste-like substance that we squeeze right into Garrett's mouth. The stuff is disgusting, but its highly concentrated formula should do the trick. Garrett remains logy but now his belly hurts as well. All we care about is raising his blood sugar. I call my brother, who says if we need to administer a glucagon shot — which causes the liver to release stored sugar into the blood — we need to use only half the vial. I hope to avoid that because those shots have side effects, such as headaches and nausea. We continue with the juice and the gel and the testing, bombarding him with as much sugar as possible, until he finally becomes overloaded. He vomits, splashing the calories all over his pajamas, couch, and carpet. We clean him up, wait awhile, and dread the prospect of having to reprime him with food and drink. We test him one more time, then have him sleep with us, Sheryl and I touching his soft skin throughout the night to detect perspiration, trembling, or any other signs of hypoglycemia. Incredibly, he wakes up a perfect 119. "He vomited just the right amount," I e-mail my brother.

The episode has no logical explanation. Garrett received his usual amount of insulin, ate his normal meals, and didn't have any unexpected physical activity. I attribute the scare to a sudden release of beta cells, a final pancreatic spasm in the twilight of his honeymoon phase, a dark joke played by his body to humble the mortals who care for him.

On Christmas Eve, Garrett is unusually defiant before his evening shot, running through the house and yelling that he doesn't want to be big and strong and he doesn't want to have diabetes. We tell him how sorry we are that we have to do this, and we give the injection, eliciting another round of protests and punches.

I realize that Garrett is fortunate compared to the diabetic children of previous generations, who did not have the benefit of multiple daily injections and home glucose monitoring. I recognize that his chances for a long, healthy life are infinitely better than those of the thousands who have walked the same path before him. But there are tradeoffs. I don't see how a child can be subjected to the on-slaught of needle sticks and finger pokes without feeling angry or bit-ter. The relentless punctures constantly discomfort but provide no gratification. They only sting or hurt. If a child can emerge from that regimen without any emotional or psychological scars, free of rancor or resentment or a vague awareness of loss, it's a miracle, nothing less.

On Christmas Day, we go to a neighborhood party, with its usual as-sortment of cookies, confections, and sweet drinks. Garrett sticks to his diet, though Amanda enjoys the treats. When we go home, the kids fall asleep and I go to my office. I soon hear Amanda going down the hall, with Sheryl following. There's some commotion, and I go upstairs. Amanda tries to get to the bathroom but arrives too late. Probably too much fruit punch. She's six. It happens. I return to my office, and Sheryl enters a few minutes later.

"Are you going to test her?" she asks.

"What?"

"Test her blood sugar?"

I tell Sheryl that Amanda has shown no symptoms and that hav-ing one accident shouldn't sound any alarms. We should worry only if Amanda has to urinate excessively over several weeks. But I realize how much Sheryl has changed from only three months earlier, now viewing her children through a diagnostic prism.

"We can test her in the morning, even though I don't think it's necessary," I say.

Sheryl goes to bed and I try to get back to work, but it's impossi-ble. Some things can't wait until morning. I go upstairs, grab the me-ter, and tell Sheryl that I'm going to test her now. Sheryl's crying. "I don't think I'll be able to handle it," she says.

I walk into Amanda's bedroom, turn on the light, and sit on the bed. She's sleeping peacefully, her wavy blond hair falling across her face. If her beta cell function were indeed impaired, the sweets at the party would have certainly elevated her blood sugar. Is there a prayer to the pancreatic gods? I take a deep breath, poke her finger, and apply the drop of blood to the test strip. The familiar countdown begins.

5-4-3-2-1.

The meter flashes 100 — the prettiest three numbers I've ever seen, as beautiful as my daughter's smile. She doesn't even stir.

I walk to our bedroom door, nonchalantly call in, "She's fine," and go back to my office.

In the spring, we fire Garrett's doctor. Actually, we like the doctor, but we make the change because we believe that Children's Hospital does not have Garrett's best interests at heart.

For a quarter of a century, I've been having blood drawn intravenously four times a year for my A1c tests. The draws vary in discomfort, depending on the nurse's skill, but having a needle plunged into your vein is never pleasant. Only recently I had discovered that some clinics use a machine — a DCA 2000 Blood Glucose Tester — that measures A1c levels with a single drop of blood. No more painful blood draws! The machine has other benefits. Normally, your blood is sent to a lab, and you — hopefully — get the result several weeks later. But the DCA 2000 produces the result in minutes so patient and physician can discuss it.

Unfortunately, when choosing Garrett's doctor, I didn't think to ask if Children's Hospital had the machine. I assumed it did, given the prestige of the hospital and my belief that it wanted to provide the best care for young diabetics. Not until our first meeting with our physician do I ask how the hospital tests A1c's, and I'm furious when the doctor says it still uses conventional blood draws. I tell him about Garrett's experience with the IV when he was diagnosed and make it clear that under no circumstance will I allow the hospital to draw blood from him for an A1c test when a less painful option is available.

"Why don't you have the DCA?" I ask.

The doctor apologizes, explains he's been trying to get the administration to buy one but can't, owing to some combination of red tape, questions about its efficacy, and the need to show its economic value.

I don't buy the efficacy argument — it's been used for years in other clinics, including my brother's in Seattle — so I ask about the economics.

He says any significant purchase must also benefit the hospital's bottom line.

"What about patient care?"

He says that's secondary, though he believes he can work these issues out and hopes the hospital will have the machine in the near future.

When we take Garrett to his next checkup three months later, we have the same conversation about the DCA. The doctor is still apologetic, but there's been no progress. I wonder if the expense of the machine is the issue, so when I get home I look it up on the Internet. The "Web price": $2,995. I suspect the hospital parking garage generates that much in a single morning.

I realize that most hospitals are financially strapped, but I cannot support one that would subject its patients to a painful, invasive procedure so it can save a few bucks.

Of course, I have my own double standards. I continue to have my blood drawn the old-fashioned way for my A1c's and have never demanded that my own doctor or hospital buy a DCA. No surprise, I guess: I want Garrett's care to be better than mine.

The blood draw isn't the only problem we have with the pediatric unit at Children's. When Garrett is close to running out of syringes, our pharmacist calls the unit to renew his prescription, but it's never refilled. Two days later I call the unit; still nothing. Finally, on a Saturday morning we're down to Garrett's last syringe, so I telephone the physician on call and explain the problem. He berates me for bothering him, saying his job is to handle medical emergencies, not prescriptions. I don't bother defending myself or saying it was his hospital that's been negligent. I just want the order called in. He does, but after I pick it up and take it home, I realize he's called in the *wrong* prescription. Garrett uses a microfine syringe with an 8-millimeter needle, but these syringes have a 12.7-millimeter needle. To a busy physician, 4.7 millimeters may not mean much, but when I give Garrett his shot, he knows immediately. He howls in pain and demands to know what's wrong with the needle.

I call the doctor back. He begins to complain again, but this time I'm pissed. I tell him he screwed up the order and to get his sorry ass on the phone and call in the right one. I hang up before he can fire back. But the damage is done. For many weeks, Garrett fears a return of the "big shot" and wants to see the needle before I use it.

In leaving Children's, we don't go far to find a new home. We cross the street to the Joslin Diabetes Center, the legacy of Elliott Joslin, which has a pediatric unit and a DCA 2000.

I've been inside the building before but now enter the pediatric area for the first time. It doesn't take long to realize why Joslin is the best at what it does — and why it loses so much money each year. Patients cherish its comprehensive services, but they are expensive. In addition to pediatric endocrinologists, nurse practitioners, and nurse educators, the clinic has child psychologists, social workers, exercise physiologists, and "care ambassadors," who make follow-up calls with patients. But our favorite is the child life specialist, a lovely young woman named Alisha who plays with the children in the activity room while the parents talk to the physician or nurse. What a great idea. Our previous appointments had always been disjointed because either Sheryl or I would have to keep Garrett entertained. Alisha also helps families develop strategies for taking shots. She gives Garrett a small football to squeeze during his injections, and it becomes a wonderful diversion, reducing his discomfort. Later I'm amazed when Alisha calls me one afternoon just to see how Garrett is doing, a modern equivalent of Elliott Joslin's "wandering nurses" who made house calls.

Few clinics, of course, have child life specialists or care ambassadors because their salaries are not covered by insurance. Joslin absorbs the losses because it puts its patients first. Indeed, the highlight of our first appointment occurs when we're directed to the house phlebotomist (itself a surprise; it's my first experience with a phlebotomist) for Garrett's A1c test. Alisha suddenly appears with a picture book, and as the phlebotomist prepares to prick Garrett's finger, Alisha turns the pages, asking about the colors and shapes, enthralling my boy. He is oblivious to the test, which shows a good A1c, 7.4. I want to hug Alisha and print up a thousand bumper stickers that read: TAKE A PHLEBOTOMIST TO LUNCH.

We get some helpful tips. Our physician suggests we use syringes with half-unit increments, which I didn't know existed but are perfect for Garrett. We set up an appointment with a nutritionist, our care ambassador gives us an information package, and our nurse educator gives us a spring-loaded injector set to help with shots. All this for a $15 co-pay.

The whole operation is smooth, professional, and caring, and I think about Elliott Joslin. Whatever misgivings I have about his puritanical rigidity or his blame-the-patient mentality, his clinic defies all the pernicious trends in health care and now benefits my son. I wish I could walk down the hall, open his office door, and thank him.

Sheryl keeps photo albums of the kids. One night I happen to look at the pictures of Garrett on our summer vacation in St. Louis, a month before his diagnosis. They capture his final prelapsarian days — laughing at the baseball park, running beneath the Gateway Arch — unaware of the betrayal brewing in his body. I consider the question that haunted me at the outset. Was it really a mistake to have Garrett? Of course not. He brings so much joy to so many people. If he were not around, the world would be a much smaller place.

CHAPTER 7

New Lows

WE SIGN GARRETT UP for indoor soccer league, and his first practice is on a bright Saturday morning in November. I pack his snacks, meter, and juice boxes (low-calorie and regular). Before leaving, I test my own blood sugar. Because even moderate hypoglycemia can impair judgment or cause nervousness or sweating, I have a whimsical guideline on blood sugars: never drive under 70, never play poker under 80, never get married under 90. This morning I'm 72, so I eat a banana and carry two packets of Skittles in my pocket.

We get into the Pilot and drive about fifteen miles to the Canton Tot Plex. The kids run drills, kick the ball up and down the field, and chase the high school girls who are their coaches. Garrett's going full bore, so at break time, I give him some regular juice and keep a close eye on any signs for lows. As the hour-long practice winds down, he's looking fine, though I detect a very slight . . . emptiness in my stomach. I think about testing with Garrett's meter, and I also consider eating something, but we'll be meeting Sheryl, Amanda, and some friends for lunch, and I don't want to snack now if I don't have to.

As we head to the car, I give Garrett some Cheerios for his morning snack. Something is wrong, but I don't know what it is. I put Garrett in his car seat and buckle him up. Then I sit in the driver's seat, put on my seat belt, and start the car. I recall seeing the red light on the dashboard signaling that Garrett's door is not shut, but I don't react to it. I wheel out of the parking lot and drive down a long back road. This is our first time at the Tot Plex, and when we reach the main street, Route 139, which leads to the highway, I don't know which way to go. Everything is vague. I think about stopping at a gas station for directions but decide to motor on. I turn left, drive down

Route 139, and see a sign for Route 24, though I don't know if I should go north or south. I pick one — I don't remember which — and accelerate onto the highway.

If low blood sugar caused your left elbow to hurt, treatment would be easy. When you felt the pain, you would eat or drink something until the pain disappeared. If your left elbow continued to hurt, you would do more of the same. However long it took, a closed-loop signaling mechanism would tell you what to do. But hypoglycemia doesn't work that way. Instead, it represents the most diabolical aspect of diabetes, impairing the one organ in your body — the brain — that you need to treat the problem.

The brain is a glucose-eating machine, requiring a continuous supply to function. As David Shenk writes, "Enormously powerful and potato-chip fragile at the same time, the brain is able to collect and retain a universe of knowledge and understanding, even wisdom, but cannot hold on to so much as a phone number once the glucose stops flowing." When that happens, the brain begins to shut down, so the body has developed warning signals. When its blood sugar drops below 60, the body releases counterregulatory hormones, including glucagon (from the pancreas), cortisol, and epinephrine (adrenaline). These hormones produce distinct "autonomic symptoms," or signals, of hypoglycemia, including shaking, perspiring, tingling, and a racing heart as your body tries to raise its own blood sugar to save itself. If that doesn't work and the glucose level continues to fall, the brain's malfunctioning accelerates, triggering neuroglycopenic symptoms that include drowsiness, confusion, fatigue, and slurred speech. If you lose consciousness, another person can administer a glucagon shot, which causes the liver to release glucose. Without one, blood sugar levels of 30 or below can progress to coma, seizures, or death. An estimated 2 to 4 percent of deaths of type 1 diabetics have been attributed to low blood sugar, many of those in car accidents. The potential for calamity deters many doctors from using insulin properly. Studies have shown that even hospitals, whose controlled environments should be ideal for aggressive treatment, often fall short. A research fellow in endocrinology at Mass. General Hospital told me why: the staff members "are reluctant to risk killing someone with insulin." This mindset led doctors to coin a maxim for the surgical ward: "Blood sugar and sex — you can never have too much." Which may be good for the surgeons but not so good for the patients.

Hypoglycemia can also occur in type 2's who use insulin, but the very nature of their disease — insulin resistance — acts as a bulwark against bad lows.

I once tried to explain to an actress — she was playing a diabetic who has an insulin reaction — what the experience is like.

I said that reactions hit you differently, depending on the type of insulin causing it. Those that come most quickly are the most dangerous. Other times, the feeling is barely perceptible and you're not sure what's happening, but it builds momentum until you recognize you're in trouble. On rare occasions, you have no idea what's about to hit, and you just lose consciousness. But typically you are aware, you feel your body sweating, your limbs trembling, your eyes losing focus.

I continued with a metaphor that I had thought about but never really expressed: "You feel like you've plunged into an ocean, your body is sinking, and you feel yourself falling farther and farther away from the surface. You know if you keep descending, you're going to run out of 'air,' and you may never come back up. Your body may be pouring out sweat, but the main thing you feel is going deeper into the water, away from the surface, and a sense of desperation is coming over you. To reverse your fall, you know you have to eat or drink something, though sometimes your liver will realize how far you've fallen and will release glucose on its own. Every second matters because the farther you go down, the darker it gets. Finally, you drink some orange juice or eat a candy bar, but that doesn't mean you're saved. It just means you've stopped falling. Now you have to kick your way to the top. It takes time for the food to be absorbed, so during this time you start kicking, and your legs get stronger and you're moving much faster until you can finally see the surface again. You've almost made it and then — *bam!* — you break through the surface; you can breathe again, and your blood sugar is fine. You're soaked with sweat, but it's a beautiful day."

I'm certain the actress never imagined her scene in that fashion, but I also realize there are limits to an outsider's understanding: if you've never been there — a hundred feet under water, and falling — you'll never really know.

Insulin's disturbing power — rendering subjects half living and half dead — prompted horrific experiments on patients with mental disorders. In the 1930s, psychiatrists noticed that hypoglycemic diabetics being treated for morphine withdrawal were no longer "restless and agitated" but had become "tranquil and accessible." A Viennese physician named Manfred Sackel intentionally induced diabetic coma in psychotic patients, particularly schizophrenics, believing that if the brain were deprived of sugar, the impaired cells would die.

Despite scant evidence of its success, the treatment gained popularity in the United States and Great Britain as well as Switzerland. One of its most famous subjects was John Nash, the math genius who would win a Nobel Prize in economics. In the 1950s Nash, suffering from schizophrenia, was admitted to the insulin unit of Trenton State Hospital in New Jersey. For six consecutive weeks he received a morning injection, inducing a drowsy, half-delirious state. By midmorning he was comatose, sinking deeper into unconsciousness until "his body would become rigid as if it were frozen solid and his fingers would be curled." A nurse would then thread a rubber tube through his nose and esophagus and administer glucose, which would slowly revive him.

The dangers in this treatment were obvious. Very often patients would have seizures, thrashing around, biting their tongues, breaking bones. Overheated patients had to be packed in ice. As one would recall, "They make me come back every day, day after day, back from nothingness . . . very little of it is clear in retrospect save the agony of emerging from shock every day." Some patients simply died, with the treatment producing an overall mortality rate of about 1 percent; by the 1960s, it had been phased out by most hospitals. John Nash remembered it as "torture."

At least the schizophrenics were in a controlled environment. What makes insulin reactions so frightening is that diabetics can still function after they have lost awareness. They are neither conscious nor unconscious but in a netherworld, functional but oblivious, placing themselves and others in mortal danger. Debra Hull, for example, decided to go to Linens 'n Things to buy a towel set for a friend's shower gift. She had been rigorously exercising to tighten her control, creating downward pressure on her blood sugar. On this day, the last thing she remembered was driving down a busy street in Chicago toward the store. She woke up in the back of an ambulance with an IV pumping dextrose into her arm. Incredibly, she had actually reached Linens and Things, walked through its doors, returned to her car, and was traveling 50 mph when she slammed into a parked truck, her head crashing into the windshield. She had no memory of any of it, and she was lucky to suffer no permanent or disabling injuries. When her husband retrieved her purse from her wrecked automobile, he found a store receipt indicating that his wife — while unconscious — had purchased a banana holder! She didn't even know such an item existed, and she could only wonder how she'd acted in the store. Her friends tried to cheer her up by bombarding her with banana magnets, banana stationery, banana barrettes, banana potholders, and, of course, bananas.

No one tracks hypoglycemia-related car accidents, so estimating their frequency is difficult. But anecdotal evidence is not hard to find. As a child, for example, Mary Slobonik was nearly killed when the car her father was driving went through a stop sign and was hit by a truck. Her father has diabetes and was presumably having a reaction at the time. Slobonik herself was diagnosed at twenty-one and was in another serious accident, this time while she was driving. She was at a clothing store near her house in Marysville, Washington, and as she later recalled, "I remember being in the store and not knowing why I was there or what I wanted to pick up. I left the store and the next thing I knew I was in the emergency room. I had driven erratically for about six blocks and left the road, crashing into a tree," puncturing a lung and bruising her ribs. She was grateful that she hadn't hit a pedestrian.

While low blood sugar rarely leads to violence, odd behavior is certainly possible. As the director of a summer Bible camp, Linda Heffelfinger, of Hicksville, Ohio, had been worrying about craft projects for the class she taught. Her son, Paul, was also at the camp. In the middle of one night, she walked into his room with a pair of scissors and pulled him out of bed. She then looked at a wall and began speaking to it as if it were a class full of children. Paul realized his mother was having a reaction and tried to talk her back to reality, but that didn't work. She then took the scissors and instructed the class to cut something — which turned out to be Paul's underwear. She cut it right off his body. Paul finally ran away, got some chocolate for his mother, and called for help. As she now says, "My poor son, I could have cut him or hurt him. We laugh about it now, but deep down no one can understand how painful this memory is."

Many of these patients have "hypoglycemic unawareness": they don't feel any symptoms until their blood sugar has descended to dangerously low levels. Such patients have passed out while teaching a class, dancing in a nightclub, or praying in church. Jonathan Sparks, of Pennsville, New Jersey, was in a Jacuzzi while on vacation when he began spitting water at his wife. Sharon knew that he needed sugar, but as he began to lose awareness, her task was complicated by two factors. Her husband weighed four hundred pounds, and they were at a nudist club. With some assistance, she pulled Jonathan out of the tub and gave him a shot of glucagon, so he had begun to recover by the time the ambulance, the paramedic, and two fire trucks arrived. "The place went nuts," Sparks says. "I guess the paramedics and the firemen had a great time looking at all the folks prancing around in their birthday suits."

Some diabetics have unconventional explanations for lows. One woman says her blood sugar falls when she shops at Marshall's. Researchers know that stress and volatile emotions — in addition to insulin levels, diet, and exercise — influence glucose levels. Severe lows cause some to have out-of-body experiences, near-death episodes, or epiphanies. Erika Parent, of Pembroke, New Hampshire, was rushed to an emergency room with a blood sugar of 10. In the ambulance, she thrashed about, crying, "It's not my time to go! It's not my time to go!" She later recalled, "I could see my hand slipping out of someone else's, and I was begging them — my children needed me." Her response surprised her because she always believed that when it was her time to die, she would not fight it. "But I did because I had to for the two most important things in the world to me — my sons," she said.

As we drive down the highway, I pass several exit signs, but I'm not sure which exit to take or how to get home. I'm frustrated because we're going to be late for lunch. I have the wherewithal to call Sheryl on my cell phone and tell her that I can't figure out where to go. I don't remember what Sheryl said, but the conversation ends and I continue to drive. The red light still indicates that Garrett's door is ajar, but I don't respond. I'm not perspiring or trembling, and I don't recall thinking that I needed to pull off the road. I turn onto another road, heading north, and hope it will lead me home. I'm in the far right lane, probably traveling about 45 mph.

I've had many reactions in my life, but very few that I couldn't treat on my own. The symptoms are usually overt and pronounced. I can feel my body trying to claw its way back to normal, at least until I can eat or drink something. But this one is different. No autonomic symptoms have surfaced, no counterregulatory hormones have been triggered, no warnings have been recognized. I continue to drive, lost. My brain is shutting down, but I don't know it. I am falling into the abyss, gently and quietly, with my son in the back and cars flying past us. All is peaceful.

Thump.

We're off the highway, grazing branches and scraping leaves. Very briefly I feel terror, but before I can react or even register what is happening, the car stops. Actually, it flips like a 4,400-pound pancake. As the police report said, the car had entered an off ramp, failed to stay in the lane, then "drove straight for approximately 300 feet until it rolled over and came to a rest on its roof in a water drainage area . . . The vehicle was not visible from the roadway."

The impact jars me back to life, though I'm now sitting upside down in the car. As I release my seat belt, Garrett yells from the back, "You have to be more careful!" He looks fine. His door never opened, and his seat belt kept him secure. The windshield is cracked and dented, but it didn't shatter. My cell phone rings — I'm sure it's Sheryl — but I can't find it. I need to call for help, but the phone is lost. (We later find it wedged between the windshield and the dashboard.) I open my door and step into the ditch filled with water. Garrett, frightened, is still reprimanding me. If he was crying, I don't remember. I lift the handle on his door. He has on his purple soccer shirt and shin guards, and I confirm that he is unharmed. So am I, save a small cut on my left pinkie. I take the Skittles out of my pocket and begin to eat them.

A woman appears. She has straight brown hair and, as she navigates her way through the muck, seems agile and athletic. "I saw you drive off the road," she tells me. She says she's a nurse and she's already called an ambulance. She asks if we're okay.

I say yes, I just need to get my son out of here.

I lift Garrett out of the car and hand him to the woman, who's on the other side of the ditch. I make my way to safe ground and retrieve Garrett, and we walk up a slight hill until we reach the highway. A state trooper is waiting for us.

The sun is out, but it's cool. My shirt is damp from perspiration, and my shoes and pants are wet and muddy as well. Garrett's not wearing a jacket, so I want him to sit in the police car to stay warm. I tell the trooper that I have type 1 diabetes, that my blood sugar went low, and that I "lost awareness" while driving. The woman describes how we drove off the road. The trooper looks at my driver's license. He's not hostile, but he does ask a battery of questions. Where are we coming from? Where are we going? Do I live with my son? I answer the questions but also say that I'm concerned that Garrett is getting cold and ask if he can sit in the police car. The trooper says no. The cut on my finger has left a trace of blood on my shirt, and he says he doesn't want any blood, from Garrett or me, in his car. I want to tell him that this is an accident, not a crime scene, but figure this is no time to start an argument.

So I continue holding Garrett in the cold until an ambulance arrives. The woman stays close to us, and I use her cell phone to leave a message for Sheryl that we've been in an accident but we're okay. The trooper tells me that he could take my license right now and declare me a threat, but he says he won't. Instead, he tries to be empathic. "I know what diabetes is all about," he says. "My mother died from it a few years ago. She was fifty-five."

I'm not sure if this is supposed to make me feel better or worse, but under the circumstances, it's more information than I need.

The woman, part stranger, part savior, remains at our side, helping me answer questions and ensuring that neither Garrett nor I need medical attention. I don't recall what prompted the gesture — perhaps in recognition of the miracle that we were unharmed — but either she gives me a little hug or I give her a little hug. The ambulance arrives, and I mildly protest going to the hospital. We're not hurt; we'd just like to go home. But we don't have a choice. We have to go in for observation. The trooper repeats that he could take away my license but instead he is only going to give me a $100 ticket for "failing to stay in marked lanes." He tells me I can appeal it.

I'm not too concerned about tickets right now. Before we get in the ambulance, I tell the woman that I want to properly thank her and ask if she has a business card. She declines to give me one or even to tell me her name, saying I don't owe her anything. I insist on wanting to do something, but again she refuses. I'm fairly confident that, even without her, Garrett and I would have made it to the highway safely, but we'd have had no way to call for help. If I had been injured or immobilized, she might have saved our lives. I later discover her last name on the police report, but she apparently didn't disclose her first name, so I still have no way of reaching her. Strange. She's willing to save my life, but she doesn't want to reveal her identity. Perhaps she doesn't trust me, or perhaps she's just a good Samaritan who feels she does not need to be compensated or recognized. I'm sure our paths will never cross again, but I'll always remember her as Garrett's guardian angel.

The medic looks at Garrett's bracelet and says, "He has diabetes also?" I figure that's one good thing about the day: we know that Garrett has the right insignia on his bracelet. We're riding in the ambulance, and they've taken my blood sugar. It's 62, and they give me a tube of glucose. We're taken to the South Shore Hospital's emergency room. It was only six weeks earlier that Garrett had been admitted to Children's Hospital, so I guess by now he's an old pro. He neither cries nor complains and is remarkably composed. I'm the one who's losing patience. A nurse wants to put an IV in me to raise my blood sugar. I tell her that an IV might be appropriate if I were unconscious, but a glass of orange juice will do the trick.

"We can't give you orange juice without a doctor's approval," the nurse says.

So I sit, still a bit shaky, and wait for authorization. I realize my blood sugar is falling and ask — indeed, plead — for juice, but the nurse says no. I wonder what would happen if I just barged into their kitchen, grabbed the juice carton, and started drinking. Could I end up in the hospital and jail on the same day? A second nurse comes in. "The boy is doing fine," the first one says, "but the father isn't cooperating."

They give me a gown so I can remove my shirt. They also test Garrett's blood sugar, which is 200, and feed him lunch. After about twenty minutes, a doctor shows up and approves the orange juice.

Sheryl soon arrives. I tell her how sorry I am for everything. She hugs both Garrett and me and says, "I wouldn't trade you guys for anything."

It was the first time I'd been in such an accident, and I knew what caused it. While my control has been consistent over the years, with A1c's ranging between the high 5s and low 6s, my exercise is sporadic, a byproduct of deadlines and other commitments. I prefer intense workouts, mostly long runs, and in recent weeks I had been pushing myself. I know to reduce my insulin and had not experienced many lows, but that can be deceptive. Your liver stores glucose in a form called glycogen, which is released when your blood sugar falls, so your body can self-correct miscalculations. Persistent exercise, however, creates two risks. First, it has a "trailer effect," not only lowering your blood sugar during your activity but later as well. Second, exercise can exhaust your glycogen supply so that when your next low comes, you have no defense.

In my case, before we left the house, my blood sugar was 72 and I ate a banana. On other days, that would have been enough, but not today, not after all my recent workouts. I also erred in not recognizing how much insulin was still in my body from my morning shot of

Humalog, which begins to take effect in fifteen minutes, peaks at about one hour, but isn't gone until four hours have passed. Even the fastest-acting insulin has a long tail, which I had failed to offset with food.

Insulin and food, food and insulin. I imagine them like armies in the night, battling inside a diabetic's body. Survival requires a balance of forces. If one become too strong (for example, if you overeat), then reinforcements are needed (you require more insulin). And if one army has retreated entirely (you skip several shots or meals), the remaining brigade of rampaging food or unchecked insulin unleashes its destructive force on the body itself, causing ketoacidosis or hypoglycemic coma. The battles never produce a winner. The armies simply live to fight another day.

I'm also reminded of Nietzsche's line: "That which does not kill me, makes me stronger." With insulin, the inverse is true: "That which makes me stronger, can also kill me."

Before we leave the hospital, the doctor pulls me aside and sternly reminds me about the seriousness of the accident. She asks if I had consumed any alcohol recently, and I told her I'd had a beer the night before. This, she tells me, is why I passed out and why I can't be so reckless in my personal habits.

The doctor is young and earnest, and I'm sure she read somewhere that alcohol can lower a diabetic's blood sugar. Of course, if I bottomed out every time I had a beer, I'd have been locked up long ago. But I can't expect an ER doctor to understand the nuances of diabetes, and I don't try to correct her mistake. Besides, I'm in no position to give her a tutorial. I just want to go home.

A tow company hauls the Pilot to its lot, and we track it down to retrieve papers from the glove compartment. I check in with the driver who rescued the car and ask if it can be salvaged. He looks at me incredulously and shakes his head.

"You were in that car?"

I nod.

"I can't believe you're alive."

Indeed, the Pilot is crushed like an accordion and caked with mud. The heft of the car probably saved our lives. Sheryl looks on the bright side: "It needed new floor mats anyway."

The hospital sends us an invoice, which our insurer covers. I received a Band-Aid for my finger and was charged $799; Garrett got a peanut butter and jelly sandwich and was billed $598.

I decide to go on the insulin pump; some studies have indicated that it decreases the risk of hypoglycemia, and I need to minimize the chances of another catastrophe. I know that a pump won't prevent a car accident. Your best bet that won't happen is to always test before driving, and even that won't guarantee safety. (Your blood sugar can drop during a long drive.) Nonetheless, my brother has been on the pump for years, and he recently showed me how newer models track how much insulin is still in your body after you "bolus,"* which is critical information in gauging which way your blood sugar is heading. I also fear I could lose my driver's license and, if that becomes an issue, I would want to show that I've taken steps to prevent another episode.

Fortunately, my license is never in jeopardy, but I do receive my $100 fine for driving outside the lines. At first I'm going to pay it, figuring it's a small price for what could have been an unimaginable tragedy. But I reconsider. Should I be held responsible, legally or morally, for what was, after all, a medical problem? Did I really violate a law when I was incapacitated? Guilt requires, if not intent, at least awareness — the ability to distinguish between right and wrong — which I obviously didn't have. I decide to appeal the ticket.

Seven weeks later, the Quincy District Court summons me for a hearing. I had intended to show the court officer my insulin pump. Though the officer is not to judge my fitness to drive, the pump might be important to allay any of his concerns. But it dies the night

---

* A "bolus" is the short-acting insulin given before eating or to correct for high blood sugar. "Bolusing" is most often used to describe insulin pump therapy — you don't take a shot but you "bolus." "Basal" insulin is the continuous, long-lasting insulin, and with pumps, your "basal rate" is the amount of medication that continually drips into your body. "Basal-bolus therapy" is now considered the gold standard for diabetes care.

before the hearing. There's no medical crisis — I can revert to injections until the manufacturer sends me a new one — but it obliterates my courtroom defense. How can I tell the officer about this technological marvel when it's broken? The answer: bluff. Just pretend that it's working. I've bluffed in poker, so why not in court? I strap it on and head to Quincy.

About thirty-five of us, all appealing tickets, jam the courtroom. The officer enters, a husky man in a business suit who clearly would rather be doing something else. He offers no pleasantries, only brusque commands. "Stand away from the door!" he barks. "If you have a cell phone, turn it off!" Perhaps he uses hostility to discourage defendants with frivolous claims. Whatever, he takes the role. I don't want my case to be called at the end — I'd be here all day — but I don't want to be near the top either. I prefer to see how the other defendants handle themselves before I have to appear.

The first case is called: "James Hirsch!"

Unbelievable. Walking to the front of the courtroom, I feel like a sacrificial lamb.

A uniformed officer reads my police report in a stentorian voice, ensuring that every misdeed is heard by the entire room. "He's a diabetic . . . he drove off the road . . . he rolled his car . . . he was taken to the hospital for observation . . . his son was in the back seat . . ."

I just stand there, feeling small and humiliated. When he finishes listing my transgressions, I'm supposed to speak, but I suddenly feel as if I have to justify myself not only to a hearing's officer but to a room full of strangers.

"Well, my feeling is that . . . As the officer said, I had a low blood sugar reaction, and my feeling is that I should not be held responsible for violating a traffic law if I was unaware of my actions at the time."

"You're diabetic and you crashed your car, and now you come in here," the officer declares, his words filling the chamber. "You could have hit someone else or injured your son. How do I know you're not going to hit someone else today? Why should you even be on the road?"

"Well, I'm now on an insulin pump, and here, I'll show it to you." I begin to pull the useless device from my pocket, but the officer stops me.

"How does it work?" he asks. "Is it attached to you?"

"Yes, it's attached. The insulin goes through a tube. You get more information and you —"

"Does it work automatically?"

"What?"

"Does it work automatically?"

"No, I have to press buttons to control how much insulin to give."

"And how do you know?"

"It's based on my blood sugar level and how many carbohydrates I eat. I also receive a basal amount of insulin, or a continuous amount, just like a pancreas."

The officer looks at his sheet, picks up a pen, and yells, "Not responsible!"

"What?" I ask.

"Not responsible," he repeats, writing "NR" on my form.

The encounter is bizarre. My insulin pump was supposed to show that I've improved my management of the disease, but it had nothing to do with the purpose of the hearing: namely, was I responsible for the actual accident? That question was ignored. I don't resent the officer's surliness. Rather, I'm dismayed at his implication that the accident was due to my indifference, as if I were cavalier about diabetes, as if I didn't spend every day of my life trying to manage it, as if blame lies only with the patient and not the disease. He made his verdict known to me and to everyone else in that courtroom. I was guilty. I was not responsible, but I was guilty.

CHAPTER 8

Dr. Bernstein's Solution

RICHARD K. BERNSTEIN is quite certain that he knows how to help a diabetic avoid hypoglycemia, maintain truly normal blood sugars, lose weight, and drastically reduce insulin requirements — all while in defiance of mainstream medical practices. Praised by some as a visionary clinician and ridiculed by others as a self-defeating martinet, Bernstein is one of the most famous diabetologists in America. He also has the country's most unusual diabetes practice, imposing burdens and responsibilities on patients not seen since the starvation era of care. Working out of his home office in Mamaroneck, New York, he often tells his patients about the superiority of his approach, which may account for his unconventional billing and administrative policies. He charges $5,500 for an initial three-day visit and $480 per hour for follow-ups. He refuses to deal with insurers, so patients pay him directly and try to recoup the costs on their own. He won't call in prescriptions. Pharmacists must call him instead.

But what most distinguishes Bernstein is his controversial message, spread through books and audiotapes that resonate with desperate patients and have made him an industry iconoclast. He believes that diabetics have been betrayed by a lazy, self-interested medical establishment; that new technologies like the insulin pump benefit manufacturers more than patients; and that the failure of the ADA to advocate lower glucose levels is part of "the rape of the diabetic."

The only successful therapy, he argues, relies on essentially the same tools that were used eighty years ago — insulin and diet (plus a glucose meter), and he invokes some of the same principles of the starvation guru, Frederick Allen. He instructs his patients to follow a radical low-carb diet, forbidding all snacks and most fruits, breads, and starches. Even water is forbidden except when thirsty. For patients who can't curb their appetites, he recommends hypnosis.

Bernstein isn't too concerned with the practical challenges that extreme diets might impose on patients. What matters is that these diets work, and he has an ideal poster boy as proof: himself. Diagnosed in 1946 at the age of twelve, he spent more than two decades foundering with chronic high blood sugars. His stunted growth produced a self-described "115-pound weakling." He had painful kidney stones, "frozen shoulders,"* high cholesterol, and impaired sensation in his feet. His frequent lows were accompanied by fatigue, headaches, and irritability. He married and became a father, but with his vision failing, his kidneys declining, and his heart worsening, he doubted that he would see any of his three young children grow up.

But everything changed in 1969 when he saw an ad for a machine that tested a patient's blood sugar in one minute with a single drop of blood. Like other diabetics at the time, he tried to monitor himself through urine tests, but he knew they were of little use in determining glucose levels. He had no reason to believe that tighter control would improve his health, but he thought a glucose meter would allow him to catch hypoglycemia earlier. His wife, a physician, ordered the machine, and he began testing his blood sugar five times a day.

Dick Bernstein was probably the first layperson to monitor glucose levels outside a hospital or doctor's office, and what he learned shocked him. His numbers swung from under 40 to more than 400, which he figured contributed to his lethargy. An engineer trained to solve problems, he began increasing his daily injections and cutting down on carbohydrates, which helped reduce the volatility in his blood sugar, but few readings fell in the normal range, and his physical ailments progressed.

He next stumbled on research that described how normal blood sugars had prevented or reversed complications in diabetic animals. Excited by this possibility, he further refined his insulin and diet to achieve near-normal glycemia, and the benefits were immediate. His serum cholesterol and triglyceride levels dropped to normal. His en-

---

* They stiffen from the thickening and contracting of the capsule surrounding the shoulder joint.

ergy rose, his muscle mass increased. The protein in his urine eventually disappeared, and his kidney function returned to normal.

Bernstein tried to publish his findings in medical journals but was rejected. Disgusted, he entered medical school himself in 1979 at the age of forty-five, published his first book on glucose control the following year, and in 1983 opened his own practice, preaching an anticarbohydrate gospel almost twenty years before the low-carb craze while positioning himself as the field's leading gadfly. His blood sugar goal is 40 percent lower than the ADA's — a goal so tight, the ADA and most medical experts believe that it creates unnecessary risks for hypoglycemia without significantly decreasing the chances of complications. Bernstein says the ADA, as well as the NIH, isn't concerned about complications or a cure. "They have a stake in diabetes," he says. "Take away diabetes, and you take away their careers."

Bernstein's strident rhetoric and uncompromising therapy appeal to patients who feel poorly served by their doctors or victimized by the health care system. Howard Steinberg was diagnosed at ten, and by the time he was in his twenties, he was looking for a new doctor. He found Bernstein. During his first visit, Steinberg mentioned the name of his current doctor — a former president of the ADA — and Bernstein responded, "He's obviously killing you." Bernstein made it clear that Steinberg's current diet, which included bagels, pizza, and orange juice, guaranteed poor control and bad outcomes. "It was a wake-up call," Steinberg says. "He didn't accept common wisdom, and that's why he's brilliant."

Bernstein helped him develop his "hyperglycemic paranoia" so that Steinberg will use unusual means — injecting insulin directly into a vein, for example — to quickly lower his blood sugar. He also reduced the carbs, and over the years his A1c's have been in the low 6s, which easily meets the ADA's recommended value of less than 7, is better than the levels of most diabetics, and has satisfied Steinberg himself. The results, however, have not pleased his mentor, who wants all A1c's in the 4s, which would match the blood sugar levels of someone without diabetes. Steinberg says he can't achieve that goal because he cheats on his diet. "I've given up being that extreme," he says. But he's still the better for trying, and as the founder of dLife, a multimedia concern focusing on diabetes, he has incorporated some of Bernstein's ideas, as well as Bernstein himself, into dLife's half-hour weekly cable show. "What I've learned from Bernstein is empowerment — he's given me the knowledge to troubleshoot and solve problems," Steinberg says. "He's broken it down to the basics."

But for other patients, Bernstein's inflexibility creates problems. Anna Smith, for example, is a forty-four-year-old fundraiser from the Seattle area who was diagnosed at ten. She's had numerous complications, including retinopathy, "frozen shoulder," and "trigger finger," which was repaired surgically. She went on the pump in 1992, which improved her control, but she still suffers from nocturnal hypoglycemia. When her husband had to call 911 after her last episode, she knew she had to make a change. In her search, she discovered *Dr. Bernstein's Diabetes Solution,* and its emphasis on reducing carbohydrates made sense to her. "I used to live on high carbs, and I always felt as though I was chasing blood sugars," she says. When she asked her endocrinologist for additional help with this new diet, he said he didn't believe in it, prompting a heated argument. So Smith and her husband flew across the country for a three-day appointment with Bernstein himself.

They enter his cluttered office on a cool, sunny day in January. The seventy-year-old doctor wears a hearing aid, pecks at the computer keyboard, and squints to read the screen, but he betrays no infirmities when he speaks. "We're going to teach you how to become a survivor," he declares.

Not everything about Bernstein is comforting. He often tests his blood sugar and is frequently low, which he attributes to an unrelated illness but is still unsettling to Smith, whose very purpose in being here is to eliminate low blood sugars. Bernstein doesn't draw blood from his fingers, which most patients do, but pricks the back of his knuckle, then licks the blood off. Before lunch (smoked trout and pecans), he injects his insulin through his dress shirt, another unusual habit. When he's talking to his patient, he turns on a tape recorder, and the recording will be used as part of his set of CD-ROMs, *Secrets to Normal Blood Sugars for Type 1 Diabetics* ($159.95, plus two books). Smith later says she felt as though Bernstein was more focused on his tape than on her.

Bernstein gives all of his patients physical exams, but the session is mostly educational. In this case, it gets off to a rocky start when he informs Smith that she must abandon her insulin pump, alleging that the manufacturers "bribe doctors" for their support. "There is no medical justification for the pump," he says. "I've never seen a well-controlled pump patient. They all have complications." He believes the scar tissue created by the infusion set (the small body patch and tube that attach to the body) prevents the insulin from absorbing, a claim that pump manufacturers deny.*

Smith resists his demand, saying her pump has improved her control — her complications reflect problems from her youth — and she refuses to remove it. "It's part of me," she says.

A more serious issue emerges when Bernstein discusses her new diet. Smith has created a diary of all the food she's eaten over the past several weeks. "I notice a lot of random eating, and that's going to have to stop," he says. He believes that snacks are a modern invention — which is to say, unnecessary — and even for patients who would cover them with insulin, they must be avoided because injected insulin can never precisely mimic a pancreas. Most foods, particularly carbs, will produce some rise in blood sugar. "Once you're on our plan, we expect you to stay with it," he tells Smith.

He instructs her to eliminate all fast-acting carbohydrates: no bread, potatoes, pasta, rice, cereal, crackers, chips, or sweet fruit. Bernstein himself hasn't had an apple since Nixon was in the White House. Also to be avoided are milk (high in sugar), tomatoes (a fruit disguised as a vegetable), and commercially prepared soups (the sugar is masked by other flavors). Even carrots must be banished.

So what can Smith eat? Bernstein urges most of his patients to eat protein, including steak, fish, chicken, seafood, cheese, and eggs. But Smith is a vegetarian, eschewing all meat as well as cheese and eggs. She is prepared to give up carbohydrates, though she says she wants to eat "flat bread."

"That must go," Bernstein says.

What's left, for Smith, are vegetables. But after she mentions that she's had problems with her digestion, Bernstein believes she has gastroparesis, a common disorder in diabetics in which food empties

---

* When I went on the pump, I experienced minimal scarring and was not aware of any absorption problems.

slowly from the stomach. It can be treated with diet, but because most vegetables digest slowly, Bernstein tells Smith that she should avoid salads. Other vegetables, including asparagus, cabbage, and broccoli, can be eaten as long as they are mashed in a blender, but that sounds more like baby food than anything Smith would enjoy eating.

Searching for other options, Bernstein pulls out samples of Morningstar Farms Veggie Breakfast Bacon and Sausage Strips, which he advocates even for meat-eaters. These too don't appeal to Smith; she usually doesn't eat processed foods, and these are packed, according to the company's Web site, with "artificial flavors from nonmeat sources." Nonetheless, she takes one bite of bacon and one of sausage and grimaces. She later says, "The bacon was hideous. It tasted like cardboard. The sausage was a bit better, but I'm not going to live on that kind of food."

Some compromises are reached. Smith sometimes drinks two glasses of wine with dinner but agrees to restrict herself to one. Bernstein, however, reprimands Smith for carrying a water bottle. He approves of water for rehydration but believes unnecessary sips contribute to kidney disease. "When you get into a habit of drinking a lot, you'll be thirsty more often," he says. Smith doesn't argue, but later says, "I thought drinking water was good for you."

Within a day of her return home, Smith receives an e-mail from the company selling Bernstein's CDs with a special offer of $99. She finds this pitch odd, given that she has just spent $5,500 for the same lessons. Including the travel expenses for her husband and her, the visit cost $10,000. She praises Bernstein for his campaign against carbohydrates, "but realistically, I don't see [his approach] for me," she says. "I don't see eating such little amounts. You want to live a long life, but I don't see how much you're living if you eat his way."

Bernstein acknowledges that Smith was a difficult case — "She's a vegetarian to a point of religion" — but conveys little sympathy for someone who doesn't follow his advice. "She's eating her old way, and her gastroparesis is getting worse. So unless there's a cure, she's going down the tubes."

Not surprisingly, Bernstein's critics liken his rigid approach to starvation therapy. One reviewer on Amazon.com called him "the Taliban of the low-carb craze." But Bernstein says he has three hundred patients and has treated or "fine-tuned" about two thousand, and he disputes any comparisons with starvation diets. "My patients are happy," he says. Indeed, some of his followers have posted their own tributes on various Web sites, and they can be found at any large gathering of diabetics. Even if they don't follow all of his advice, they are familiar with his writings.

Bernstein makes some valid points that any patient would be wise to follow. He invokes "the laws of small numbers" — big inputs make big mistakes; small inputs make small mistakes. In diabetes, that means the larger the dose of insulin, the greater the chance of hypoglycemia. Like Elliott Joslin, Bernstein doesn't believe that patients should rely on "insulin stilts," recommending that a single injection should never exceed seven units. (Some patients take more than a hundred units.) Dramatically scaling back a patient's insulin clearly reduces the risk of lows, and that is only possible by following an austere diet. Even if a patient cannot achieve Bernstein's Spartan ideal, any decrease in insulin that doesn't compromise overall blood sugar levels is still an improvement.

The laws of small numbers also apply to food: the more you eat, the more difficult maintaining your control will be. That's true even if you adjust your medication based on your carbs, because you can rarely count carbs with precision. A single pear, Bernstein notes, is a good example. What type of pear is it? How much does it weigh? How long did it ripen? What orchard did it come from? You need to answer those detailed questions to accurately cover that pear with insulin. Mistakes are inevitable, which can lead to temporary spikes or sharp lows. The only way to minimize those swings — in Bernstein's view, the principal salvation for any diabetic — is to shun fast-acting carbs.

He insists that his diet is possible for most diabetics once they understand the stakes. "It's life or death," he says. But his book suggests recipes that are comically impractical. Consider breakfast. For most families, it's controlled chaos: parents try to dress and feed their children before rushing them to school while preparing for their own workday. But if you can't eat waffles, cereal, or toast — and don't want trout — Bernstein recommends such items as "French bran toast," which requires you to soak two Bran-a-Crisp crackers in water for five minutes. Mix eggs and cream with cinnamon, nutmeg, vanilla, and other flavor extracts. Place softened wafers in egg mixture for one to two minutes. Heat nonstick skillet until water droplets skitter across. Add oil to skillet and spread it around with a folded paper towel. Place wafers in pan and cook for about three minutes per side. Remove from pan and pour melted butter over them.

Such dishes may be tasty and have few carbohydrates, but whether it's acorn squash bisque, curried chicken salad with jicama, or panfried okra with tamari scallion glaze — all recommended by Bernstein — they require too much time and finesse for most diabetics, or anyone else, in a premade, preheated fast-food nation.

His requirements for children seem equally unrealistic, if not callous, for he prohibits them from eating snacks as well. Most kids, diabetic or otherwise, eat at least two a snacks a day, and preschools and afterschool programs include them as part of their daily routine. Diabetic children can cover those calories either by "bolusing" insulin through a pump or, more conventionally, by administering insulin that "peaks" hours after the injection. But Bernstein believes that snacking negates the kind of tight control he considers necessary for survival. Asked about the social consequences of a four-year-old being denied food while his peers munch on pretzels or animal crackers, he says that those children have other options: "Give them sugar-free Jell-O."

Such comments, for me, make Bernstein difficult to like or even respect, and I assume that he will end his career as he began it — alone, in a fit of pique, lashing out at the patients and peers who don't see the infallibility of his wisdom. But it's also true that his anger is his most compelling trait, connecting him with patients who feel similarly betrayed and offering them a lifeline of hope. When I'm in his office, he receives a telephone call late in the afternoon, and the caller, from Birmingham, Alabama, is put on the speakerphone. She's been crying and her voice is cracking, but she's grateful to speak to Bernstein.

"I've been studying your book for a year, and I think I've figured out the problem," she says. "But I can't find a doctor."

Bernstein says he can't treat her over the telephone.

"That's okay," she says. "Can you give me a referral for someone in Birmingham?"

"I'm sorry, but I don't know any doctors in Birmingham."

The woman pauses, then asks: "How about anywhere in the South?"

In 1921 Fred Banting, a Canadian surgeon, believed that a pancreatic extract could normalize blood sugars in diabetics. Banting (right) is seen with his student assistant, Charles Best, and their research dog, Marjorie, who was the first animal to receive insulin. *Courtesy of Eli Lilly & Company*

Relying on crude equipment, Banting, Best, and other researchers at the University of Toronto used "bathroom chemistry" to develop the life-saving solution. *Courtesy of Eli Lilly & Company*

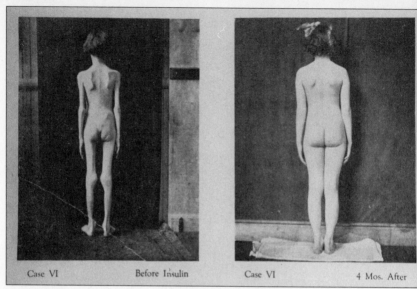

Case VI   Before Insulin   Case VI   4 Mos. After

Widely published photographs demonstrated insulin's miraculous power to literally restor*e* flesh to bones. *Courtesy of Eli Lilly & Company*

Born in 1907, Elizabeth Evans Hughes was the daughter of one of the most prominent men in America, but she herself became famous as the poster girl of the new "diabetic cure." She later married, had children, and traveled the world, but she concealed her disease at all costs, even from her own children. *Courtesy of the University of Toronto, Thomas Fisher Rare Book Library*

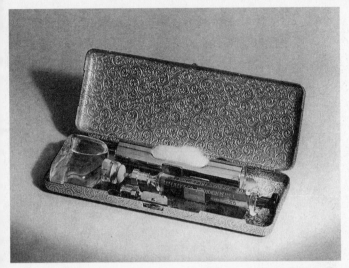

Requiring patients to administer their own injections was unprecedented, time-consuming, and often painful; the early glass syringes used large needles that had to be sterilized in boiling water and sharpened on a whetstone. *Courtesy of Eli Lilly & Company*

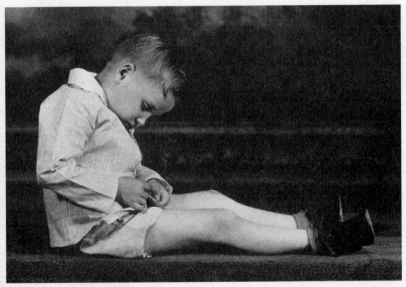

That diabetes affects children has always given the disease a special poignancy, and few images evoke more sympathy than that of a youngster giving himself an injection. *Courtesy of Eli Lilly & Company*

Our favorite picture of insulin

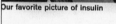

These ads for insulin, produced for Eli Lilly, reinforced the drug as an elixir for a once-fatal disease without acknowledging its limitations. For decades, the notion of diabetes as a scientific success story fueled misperceptions about the disorder. *Courtesy of Eli Lilly & Company*

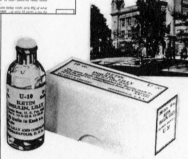

Medical Building at the University of Toronto, where Dr. F. G. Banting studied and later engaged in research that led to the discovery of Insulin.

I N S U L I N

"An Epoch in the History of Medicine.

A Boon to the Human Race."

The Lilly Research Laboratories, source of the first commercial Insulin available in the United States.

During World War II, after Shanghai could no longer import insulin, Eva Saxl relied on her husband to keep her alive. She is an extreme example of all diabetics who have survived through resourcefulness, courage, and love.

As America's preeminent diabetes clinician in the twentieth century, Elliott Joslin brilliantly recognized the importance of near-normal blood sugars in reducing the risk of complications, but his criticism of patients who fell short legitimized diabetic self-blame. *Courtesy of Eli Lilly & Company*

Representing Virginia in the 1998 Miss America pageant, Nicole Johnson removed her insulin pump for the swimsuit competition and chose to wear a two-piece suit that revealed the tape for her infusion set. She still won the pageant. *Courtesy of the Miss America Organization*

After his infant son was diagnosed, Douglas Melton redirected his science lab to beta cell research, which thrust him into the center of a bitter national debate about science, religion, and morality. *Jonathan Kannair (Kannair.com)*

In 2002 Gary Kleiman received an islet cell transplant, which improved his control but did not cure him. He calls the transplant "an imperfect but amazing therapy." *Photograph by Rhoda Baer*

Known as the Diabetes Queen of Georgetown County, South Carolina, Florene Linnen is trying to stem the type 2 epidemic by imploring residents to demand more from health care providers. *James S. Hirsch*

Though she remains the most controversial researcher in the field, Denise Faustman's central findings — curing autoimmune diabetic mice through regeneration — have now been largely confirmed. But she is many years away from knowing whether her research is relevant to humans. *Porter Gifford*

Garrett's diagnosis did not keep him from playing soccer on his first Saturday home. *James S. Hirsch*

Amanda has appointed herself the unofficial custodian of Garrett's blood sugar scores.
© 2006 Bobbie Bush Photography.

CHAPTER 9

High-tech Tradeoffs

IF THE CONVENTIONAL TREATMENT of medication and diet has failed to control diabetes, technology has offered the other great hope. Insulin, in its own way, was a futuristic drug, and even if it wasn't the cure, its discovery raised expectations that further breakthroughs would continually improve care. In fact, they have, giving patients access to designer insulins, continuous glucose monitors, prime pharmaceuticals, and daring surgical transplants. These have allowed diabetics to more readily achieve near-normal glycemia, lose weight, treat complications, and prevent coronary disease. But they have come at a cost. Many of these innovations are expensive, complicated, or not widely available. Some new treatments are so sophisticated or burdensome that they're unrealistic for most patients, widening the gap between the diabetic haves and have-nots.

This high price is evident in commercial insulin. From the outset, its origins from cows and pigs created problems: its impurities caused disfigurement and swelling, leaving children with bloated thighs and distended arms. Some patients also had allergic reactions, and the product's reliance on animal pancreata raised fears that declines in livestock could create shortages. Those fears were allayed in the 1980s by the development of genetically engineered, biosynthetic human insulin. Its purity was improved, virtually eliminating allergies, and there were no more concerns about supply.

But its effectiveness was still limited, because it poorly replicated the timing action of insulin that the body naturally produced. "Regular" insulin, derived from animals or labs, did not take effect until thirty to sixty minutes after its injection, peaked two to three hours later, and remained active for up to ten hours. If taken right before a meal, a patient's blood sugar would spike before it took effect; and by lingering in the body for so long, it might overlap other injections, increasing the risk of hypoglycemia. The treatment, in short, did not match a nondiabetic's insulin production.

In the 1990s, insulins that more closely matched that production were developed, again through biosynthetic technology. Known as insulin analogs, their protein structures were tailored to achieve specific therapeutic goals. The fast-acting analogs, lispro and aspart (Humalog and Novolog), take effect within fifteen minutes of injection, peak thirty to ninety minutes later, and last up to six hours. The long-lasting analog, glargine (Lantus), provides a steady "basal rate" for up to twenty-four hours. A regimen of Humalog or Novolog before each meal or snack, plus a daily shot of Lantus, more closely mimics the insulin release of a healthy pancreas.

But these better insulins cost more than twice the insulins they replaced, forcing patients to spend more either through direct purchases or higher co-payments — or to forgo them entirely. Vials of Novolog and Humalog cost $78.95 and $65.99, respectively; Humulin Regular, $32.23. A vial of Lantus costs $60.34; Humulin Lente, $32.23.* (In most cases, the vials are supposed to be discarded about thirty days once they've been opened.) The improved insulins are not only too expensive for some patients but for some hospitals as well.**

Indeed, some of the most exciting new diabetes drugs are so expensive that they are beyond the reach of many who can benefit. Exenatide, for example, is an injectable drug that type 2 diabetics can administer in conjunction with oral agents. Under the name Byetta, it replicates the action of a hormone called glucagon-like peptide-1, which is released in the gut and promotes insulin secretion by the pancreas. Studies show that exenatide lowers A1c levels, minimizes the sharp spikes after meals, and contributes to weight loss as well — a drug that could benefit many type 2 patients. But in pharmacies it costs between $7.14 and $8.38 a day, depending on the dose. While

---

* Prices from www.drugstore.com.
** My brother discovered this at an Endocrine Fellows Foundation meeting, in which about one-quarter of the fellows had no experience with insulin analogs because their use was denied at the hospitals where they trained.

that may be reasonable for an episodic illness, it's quite expensive for a chronic disease — and more than twice the cost of most other diabetes drugs. Nonetheless, driven in part by nondiabetics who want to lose weight, Byetta's initial sales have been outstanding: in a three-month period, starting in November 2005, monthly prescriptions rose almost 40 percent, to near 100,000. But the price, which is likely to increase (drug prices usually start low and bump up), excludes many patients. As my brother, drawing on his own clinical experience, wrote in the trade journal *DOC News,* Byetta "will break our current standards for moderate diabetes drug costs . . . This 'concierge agent' appears to have been developed for people with good insurance, not those from the typical lower socioeconomic population with type 2 diabetes, such as Medicare or Medicaid patients, those who obtain their drugs from Veteran Affairs hospitals, the uninsured, the underinsured, or those with restrictive formulary status." Unless new ways are found to make these drugs available, he wrote, "simply adding a pill or a shot will have little effect."

For insulin-dependent diabetics, insulin itself probably cannot be made any more pure or reliable, so technology holds the key to improved therapies — specifically, improvements in continuous glucose monitors and self-administered insulin pumps. The goal is not simply to improve the daily management of the disease. It is to create an artificial, or biomechanical, pancreas, in which a glucose sensor signals an insulin pump how much medication to release. By some, it is considered the best hope short of a cure, because the patient would not have to do anything, relying instead on a closed-loop system between the sensor and the pump that mimics a healthy pancreas. There is reason for optimism. Progress in insulin pumps, glucose monitors, and sensors has already improved the lives of thousands of diabetics, and medical device companies are investing heavily in biocompatible polymer implants, medical telemetry devices, and computer software. But this technology always involves costs, risks, tradeoffs, and unknowns and should not be confused with a cure-all — or, for that matter, a cure.

The limits of technology can be seen in even the most conspicuous breakthroughs. Consider glucose meters. Introduced in the 1980s, they represented the most important therapeutic advance since the discovery of insulin, and they continue to improve: they fit in the palm of your hand, deliver results in less than five seconds, and require the tiniest drop of blood. They have vast memory banks that record several weeks of data and compute average glucose levels. Their results can be downloaded and graphically displayed on a computer screen. They can measure ketones as well as glucose, and patients can store reams of information that affect blood sugar (exercise, carbohydrates, menstrual cycles); that pertain to their daily care (insulin types and doses or oral medication in half-pill increments); or that reflect their overall health (A1c levels, presence of microalbumin, cholesterol and blood pressure counts). Even doctor's appointments can be stored.

While these features look great in the ads, it's unlikely that many people actually use them. In fact, home monitoring is all but ignored by most patients: an estimated 13 percent of the (at least) 15 million known diabetics in America don't test at all; the known diabetics test on average only 1.5 times a day, according to Close Concerns, a diabetes consulting and publishing firm.

Some doctors are inclined to blame the patients. At the 2006 ADA Diabetes Expo in Boston, Lori Laffel, the chief of the pediatric and adolescent section at Joslin, gives a talk about diabetes care and lists the obstacles to home glucose monitoring. They include that patients find it too painful, inconvenient, intrusive, or expensive. I raise my hand.

"Is it possible," I ask, "that patients don't test because their doctors don't want to spend the time, money, and effort to teach them?"

Laffel agrees that doctors bear responsibility but says the problem is much larger: physicians are not adequately reimbursed to provide the kind of education that patients need to learn how to use the meters, which includes interpreting the numbers. "Do you have any suggestions?" she asks me.

I'm tempted to suggest that her presentations not make patients feel that they are exclusively responsible for their own shortcomings, but that would be unfair. Her clinic does spend the time and money to educate patients. Instead, I make a generic suggestion on which we can all agree.

"Reform the system," I say.

The meter manufacturers seem equally unaware of the gap between the technology in the lab and its application in the field. Paige Reddan, a certified diabetes educator, worked with patients in hospitals until she took a job with LifeScan, one of the leading meter manufacturers. She was surprised at how little the company understood its market. "They talked about the next round, the next product, the next product," she recalls. "I said, 'This is nuts. [The patients] don't know what to do with the last one.'"

After Garrett's diagnosis, I decide to upgrade our meters. We have the OneTouch from LifeScan but are wasting strips because we can't always squeeze enough blood out of Garrett's finger to activate the machine. It just gives us error messages, which means we have to prick Garrett yet again. So we buy the heavily advertised FreeStyle Flash, which requires less blood and also has a light, which is useful for testing when he's asleep.

As advertised, the meter functions with less blood. In fact, it doesn't begin its countdown until sufficient blood is on the strip, so we know when more is required. But the strip's hypersensitivity creates a different problem. One day I test Garrett before lunch, and he's 483. I know that can't be right. In the past, we could clean off Garrett's fingers with a tissue, but now I take an alcohol wipe and scrub them down. I test again and he's 112. But the problem recurs, sometimes even after I wipe his fingers with alcohol: 387, 412, even HI. (Again, they can't afford to spell HIGH.) I learn to double-check or even triple-check unusual numbers. At a diabetes trade show, I share my frustrations with a representative of the manufacturer, Abbott Laboratories. He assures me that the strips have been rigorously tested but says they may produce inaccurate numbers if the blood is not evenly applied or the finger is not clean. I think this means we're at fault. While the cost of FreeStyle Flash strips is the same as those for my other meter — a dollar a strip — for some reason I have a 70 percent higher co-pay through my insurer. These are the kinds of tradeoffs in most medical technologies: while my new meter requires less blood and lights up, it's less reliable and more expensive. We aren't sure if we should go back to our old meter or keep the new one — or wait for a third meter with a new set of benefits and drawbacks.

While less prominent, the insulin pump has also dramatically changed care. It's always had several attractions, including the most obvious: it relieves patients of having to take injections. At its most basic level, it provides a continuous drip, or basal rate, through a plastic tube inserted beneath the skin with a needle. (The needle remained in the body on the early pumps.) With the press of a button, patients can give themselves a bolus of insulin before any meal or snack, just as a pancreas releases insulin in response to rising glucose levels. "Pumpers" can eat whenever they want — or, for that matter, not eat — and they can more easily compensate for dietary indiscretions. An unanticipated brownie, for example, can now be covered with an additional bolus.

The first pumps, in the 1960s, were mounted on a backpack, which highlighted one of their principal shortcomings. Even if the pump worked, most of its potential customers wanted to hide their condition, not advertise it with a device hanging from their body. Improvements came gradually. By the early 1980s, a pump could be attached to a belt, but it still weighed more than a pound, had flashing red lights, and was nicknamed "the blue brick." Over the next two decades, the device got smaller, safer, more functional, more durable, and less uncomfortable, one improvement being the first infusion set that replaced the needle with a soft cannula that rested just below the skin. Another was the "quick release" feature, allowing users to temporarily detach from the pump before, say, taking a shower or going through a metal detector. Current models, the size of a pager, are high-tech marvels: they have electronic memories, insulin sensitivity ratios, remote controls, safety alarms that beep and vibrate, a waterproof casing, adjustable basal rates, and dual-wave or square-wave bolusing, which allows users to spread the delivery of insulin over time and is helpful when eating high-fat meals.

The device received a marketing break in 1998 when Nicole Johnson entered the Miss America pageant, representing Virginia and wearing a MiniMed pump. After she was featured in a local newspaper, several contestants approached her to apologize, saying they had treated her poorly because they had thought her pump was a pager and that — in their view — she had been too absorbed with getting her messages. During the pageant, Johnson concealed the contraption by strapping it to her inner thigh. She detached it, however, for the swimsuit competition; she wore a two-piece number but did not remove her infusion set, revealing what appeared to be a piece of tape on her stomach, thereby thrusting a flaw, a personal defect, in the faces of the judges. It was, she later said, an act of defiance: when she had been diagnosed five years earlier, doctors had told her she'd have to give up competing in pageants. Instead, she was crowned Miss America, and after she walked down the runway, the first woman to embrace her was a tearful Miss New York — who was also wearing a pump. "It was a victory for both of us," said Johnson, who used her celebrity to become one of America's most prominent advocates for diabetes care in general and pump therapy in particular.

While pumps now fit in a pocket or attach to a belt, they can still be cumbersome, particularly for women in fitted clothes, forcing creative solutions. Some women place the device in their cleavage — one says it's "like having a small truck between your boobs" — and they rely on a remote control to activate it. When one patient thought a man was looking too intently at her breasts, she beeped her pump. "They always do that when a guy gets too close," she told him. And pumpers have found other ways to exploit the machine. When a teenager wanted to end a conversation with her grandmother gracefully, she sounded the alarm on her pump and said her cell phone was ringing. Even a small pump cannot be concealed if you wear it, as one woman does, to a nudist beach in New Jersey. It hangs from a belt, which is okay as long as you don't jiggle when you bolus.

Pumpers can still play sports, preferably with no contact. Jason Johnson, a pitcher for the Boston Red Sox, received permission from Major League Baseball to wear his pump on the mound. As long as he doesn't slide, he shouldn't have a problem. For many pumpers, the device is like another appendage. When Melissa Hogan-Watts, of St. Louis, was on vacation, she stepped out of the shower, felt the room spinning, wrapped herself in a towel, grabbed her pump, and collapsed. It turned out that she had a bladder infection that had moved into her kidney. Though she had no memory of what happened next, she began yelling, "Don't take my pump! Don't take my pump!" She still had it in when she woke up in the hospital.

Alas, several women have had problems when their husbands demanded that they remove their pumps before sex. Laura L. Smith, of Garner, North Carolina, refused. "It's my lifeline," she says. She got divorced and married a man who was less easily distracted.

The pump hadn't interested me because I didn't really mind the shots and didn't like the thought of being attached to a device. But I reconsidered after the accident because, well, I figured I'd better do something. I'm fortunate that my insurer covers all costs for the machine ($6,000) and supplies (at least $1,500 a year). MiniMed is the market leader, but I bought an Animas IR 1200, not because I thought it's a better product — the variations, I believe, are minor — but Animas is a smaller company that relies on customer service to compensate for a less prominent brand name. Since I have trouble figuring out the remote control for the television, I want as much assistance as possible.

Many pumpers have told me that it changed their lives, allowing them to feel that they were controlling the disease instead of the disease controlling them. They bolus whenever they eat, delivering fractional amounts even for a few kernels of popcorn. Such micromanagement is an asset, but the pump's real power lies in its information technology. In fact, I consider it as much a high-priced consultant as an insulin-delivery system. For example, it tells you your IOB — insulin on board — which is important to prevent hypoglycemia. Fast-acting insulin does not hit your bloodstream like a tidal wave. Rather, it is absorbed quickly but incrementally and lingers for four hours. Many diabetics don't appreciate this lag time, and they bolus additional amounts to correct a high blood sugar but end up "stacking insulin," causing lows. The pump keeps track of your IOB, instructing you how much to give for a correction or how much to deliver before a meal, based on your current glucose level and the carbohydrates you're about to consume.

The pump, however, doesn't think on its own. You have to give it the correct information to get the right guidance. If you give it wrong information, you'll be worse off than if you didn't have it. In this sense, the pump does not make managing diabetes less onerous or less time-consuming. It makes care more complicated and forces you to become more involved with your own management, impelling you to test and adjust constantly; the device works best with highly motivated patients who have the time, resources, and wherewithal to learn how to use it.

I chose a black pump, and after a few training sessions, I was excited to start. I wouldn't call a contraption attached to your body with a plastic tube liberating, but it's nice not having to carry an insulin pen, bottles, or syringes. Patients have lost life-sustaining paraphernalia on honeymoons, business trips, and camp-outs; one woman dropped her insulin pen down a toilet on an Amtrak train and could not retrieve it. The pump also eliminates the ordeal of giving injections in public: the puzzled stares, the scornful frowns, the search for places to hide. While I've heard about diabetics giving themselves shots under unusual circumstances — such as riding down a highway on the back of a Harley — I believe I have a truly bizarre story. On the first day that I moved to New York in 1986, I found myself in midtown Manhattan at dinnertime. I was using syringes and insulin bottles and needed privacy, so I tried finding a restroom in a restaurant. But without buying a meal, I was denied access. So I went into a peep club, entered a private booth, and with a naked woman dancing through the hole, injected my insulin. I even left a tip.

Every pumper has some paranoia about how things can go wrong, though I tend to view disasters cinematically. I imagine two pumpers meeting on a train and having an affair, and the woman, a femme fatale, secretly swapping pumps or reprogramming her victim's, leading to hypoglycemic death. The movie's title? *Basal Instinct.*

A more realistic concern is that my machine will malfunction and stop delivering the drug. Without long-lasting insulin in my body, I rely on that basal drip to maintain control, which I keep in mind as I prepare for my first business trip since buying the pump. I pack my Humalog in case I need to refill the pump, but I also take a bottle of Lantus and several syringes in case the device breaks down. I also carry two extra infusion sets, two extra insulin cartridges, an extra lithium battery, alcohol swabs, a glucose meter, a lancet, strips, and snacks. And my trip is only for two days.

I take a cab to the airport, and I feel as though I might be low. I eat some raisins. I'm still not sure, so I eat a Kudos bar. But now, am I too high? No problem. I pump in 0.5 units. I would never give an injection of less than one unit, but now I think nothing of it. The pump is flexible, easy, and discreet. I test at the airport and I'm 92. On the airplane, I bolus 0.5 units before I eat a bag of pretzels. Now I feel low again, so I eat one more bag. After landing, I'm 95. Excellent. I have a layover, so I eat dinner, trying to avoid heavy carbs: coleslaw, salad, meatloaf. But it's a huge slab of meat, with Creole sauce. How many carbs? I have no idea, and I give 4.55 units. But that doesn't seem like enough. The salad has honey mustard dressing . . . and maybe croutons! I pull out my pump, press the buttons, and give one more unit.

Testing, bolusing, eating. Eating, bolusing, testing. I test an hour after dinner, and I'm 99. *Phew!* Such micromanagement was never possible, or at least practical, with the blunt instrument of injections. But this approach also creates new pressures. If perfection is possible, then imperfection is unacceptable, a betrayal of the technology. Some diabetics test their blood sugar fifteen, eighteen, twenty times a day in pursuit of a glycemic grail that borders on obsession.

At the airport, I keep my pump on when I go through security, and the metal detector sounds the alarm. A security official, a courtly African American who's probably in his late fifties, begins the patdown. I take out my pump and tell him what it is. He holds it in his hands, turns it over, and asks how it works. He seems genuinely interested, so I describe the functions and explain how it helps patients.

"My wife has diabetes," he says. Or at least that's what I think he said.

"How is she doing?" I ask.

"She's with her Maker in heaven," he says, handing me back the pump.

I can only say I'm sorry.

Pump therapy, for the most part, goes smoothly; the basic technology isn't difficult to master. I don't experience any significant lows, which is the main reason I got the pump, but I don't think that has as much to do with the device as with the increased number of times I'm testing and adjusting doses — which I could have done without the machine.

And my concerns about the pump are borne out: it dies on me, not once, but twice, in the first ten months. Animas promptly ships me replacements at no charge, and with syringes and insulin on hand, my health is never compromised. Animas lives up to its reputation for service; its representatives are unfailingly helpful. But there is still something tenuous about the entire enterprise. No matter how reliable, durable, and sophisticated the machines are, they will always be fallible, each new feature adding more complexity and another possibility for error. These thoughts occur to me in the dead of night when my alarm sounds, signaling that the insulin isn't reaching its destination — a kinked tube, perhaps, or a more serious mechanical defect. The machine reads OCCLUSION ERROR when a bolus fails and requires troubleshooting; it soon becomes a dreaded phrase.

Patient error also exists with anything that needs to be programmed daily. Some mistakes are serious, others comic. After wearing the pump for a few days, I hear the alarm and pull it out of my pocket, but there's no indication of what's wrong. I call the Animas help desk, and the woman has me run a few checks. The alarm rings again. "Mr. Hirsch," she says, "I think that's your cell phone."

On another occasion, I'm down to my last unit of insulin when I'm about to board an airplane in Philadelphia. I don't want to change my infusion set in cramped quarters, so I hastily try to do it outside the gate. But while I'm in midstream, the agent calls my name and tells me to board immediately. I stuff everything in my carry-on and reach my seat, but when I try to reassemble the pump, I discover I'm missing the small plastic cartridge cap — without it the machine doesn't work. We're already on the runway, so now I'm stuck, with no insulin, until we reach Boston and I can drive home; it's two very long hours. As soon as the plane lands, I call Animas and ask that it Fedex a new cap, but it's too late — no packages can be sent out until the following day, which means at least thirty-six hours without a pump. At least I can use Garrett's long-lasting insulin, so I can get by with three shots a day. Other pumpers who've suffered breakdowns have had to take shots every two hours to maintain their basal rate.

My pump, now useless, becomes a weird annoyance. It's programmed to beep and vibrate if it's not pumping insulin. I could deactivate it by removing the battery, but then I'd have to reprogram it.

I just put it down, and every hour or so we hear it squeal and growl. We try to quarantine it, but it's still too loud. The kids complain. So does Sheryl. When they go to bed, I have to take it with me to my office, and there it sits, keening like an abandoned animal long into the night.

The market for pumps remains relatively small, with an estimated 160,000 patients in the United States, the greatest increase in recent years occurring among adolescents. While various studies indicate that pumps produce better glycemic control and reduce hypoglycemia, it's unlikely they will ever become the principal form of care for insulin-dependent diabetics. They require more time — and often money — than conventional therapy, the fear of a breakdown creating additional stress. More important, health care providers are not compensated for the additional time a pump requires, so they have little incentive to push its use. Many providers don't understand conventional insulin therapy and would be at sea with a pump, and even the experts haven't mastered the more sophisticated technology. A teenager in Boston who went on the pump was dismayed when she started gaining weight. She attributed the problem to her pump, but her physician, a pediatric endocrinologist, said she had an eating disorder and recommended therapy. The patient then spoke to her nurse, who discovered that the physician had set the pump's basal rates too high, forcing the girl to eat additional food to stabilize her blood sugar.

My immersion into high-tech care deepens when I agree to participate in a study with MiniMed's Continuous Glucose Monitoring System, which the company believes is an important step in the evolution of an artificial pancreas. Using a spring-loaded gun, the patient inserts a sensor just under the skin of the abdomen or buttock. The sensor itself, a platinum-plated electrode inside a permeable membrane, sends a glucose measurement to a monitor every ten seconds. The four-ounce monitor — connected to the sensor by a cable and worn on your belt — records an average glucose reading every five minutes. It doesn't provide real-time numbers, but after several days the data can be downloaded to reveal continual blood sugars. Only available through a doctor's office, the system is typically worn for three consecutive days to analyze a patient's control.

Massachusetts General Hospital asks me and other volunteers to use the system for five days a week over three months, with the data used to interpret A1c values more accurately. I've volunteered for numerous studies over the years, and this one interests me as a reference point toward the goal of an artificial pancreas, which at minimum requires an implantable sensor. I also want to see how my blood sugars fluctuate on a continuous basis.

The sensor is inserted with a long needle, much longer than that on an insulin pump. Cocked inside a red plastic gun and triggered with the press of a button, it barrels through your skin. According to MiniMed, "the patented insertion device makes placing the sensor virtually painless for your patients." I've learned this much about medical ads and promotions: beware of the phrase "virtually painless." I make around forty insertions over three months, and they range from "mildly uncomfortable" to "it hurt like hell." Sometimes the needle causes bruising and discoloration. Other times, blood bubbles up out of my skin like oil from a well. In fairness to MiniMed, patients are supposed to use the system only once, for three days. Nevertheless, few invasive therapies for diabetes are "virtually painless."

The monitor is larger and heavier than the insulin pump, and the cable is thicker and more unwieldy. Sometimes the tube on my insulin pump catches on a doorknob and yanks hard on my infusion set. Now I have a second tube coming out of my body, placing me in double jeopardy. While the pump can be detached, the monitor cannot be disengaged from the sensor, which creates problems. For one thing, the monitor isn't waterproof, so MiniMed provides plastic bags to protect it while taking a shower or bath. The bag has a neck strap, which is supposed to allow me to use both hands to wash. Before my first shower, I seal the monitor in the bag, put the bag around my neck, and begin to wash. Moments later, *crash!* The monitor slips through the bag, smashes on the wet floor, and almost rips the sensor out of my stomach. I reassemble myself and try again. Again the bag doesn't hold, though this time I catch the monitor before it crashes. As the days pass, I try different ways to seal or secure the bag, but nothing works. I complain to the study coordinator, who is sympathetic but can offer no other options. I'm puzzled. Why does MiniMed spend millions of dollars to develop a Continuous Glucose Monitor, but it can't provide a lousy plastic bag to keep it dry. So for three months I take one-handed showers, using the other to hold the bag.

With all the hardware on my body, I recall Jerry Lewis's joking about a medical device for his heart. It works great, he says, except when he presses the button, his garage door goes up. At home one night, I bend down and the horn starts honking from our car in the garage. Is it a burglar? No, it's my continuous glucose monitor pressing against my electronic keychain, activating the car's alarm. At least the garage door doesn't go up.

I calibrate the monitor by testing my blood sugar at least four times a day and programming the numbers into the machine, which is also receiving glucose numbers from the sensor. If the programmed numbers differ significantly from the sensor numbers, the monitor sounds an alarm. The sensor is to last three days, but on my second night, I hear a beep after I calibrate, and the monitor reads: CAL ERROR. I call the MiniMed help desk and am told that the sensor is no longer functioning — it may have been damaged in the shower — and needs to be replaced. My next sensor, however, lasts only about fourteen hours before it beeps. I call back the help desk and am told again to change sensors, but my luck doesn't improve. The new one also lasts less than a day. The MiniMed helpers suspect that if I'm not destroying the device in the shower, I may just have bad sensors, so they send me a new batch in a refrigerated package. No matter. The beeps keep coming, waking me up at night, interrupting my work, and confusing me. When I beep, I don't know if it's my pump or my monitor (at this point, I've ruled out the cell phone). It's even worse in the car, because I can't tell if the beep is coming from one of my medical gadgets or the dashboard. When my brother comes to visit, he's wearing a MiniMed sensor that provides real-time glucose data and is in clinical trial for FDA approval. We're driving and a beep goes off. "Is that you or me?" I ask him.

At some point, a MiniMed rep tells me that sensors aren't supposed to last three days — which is not what the company had told the study coordinator. Rather, they can last *up to* three days, so malfunctions on the first or second day are not necessarily unusual. The variance underscores a basic challenge with the technology — or, for that matter, any effort to implant an object in the body. Once the sensor is beneath the skin, the body tries to protect itself against foreign material by encapsulating it with cells, changing the sensor surface. Researchers trying to develop an artificial pancreas have to eliminate "sensor drift," which would cripple the entire system.

The bulkiness of the device is another problem. While I can jog with my insulin pump, the monitor and the delicate sensor make such an activity impractical. There are psychological issues as well. I'm not particularly self-conscious about having diabetes, but one day I'm standing shirtless, my pump in one hand and my monitor in the other, the tubes looping across my body. Sheryl walks in and forces a slight grin. I feel embarrassed and think about apologizing but say nothing.

Self-image matters. I know of a patient who stopped wearing his pump when he went to college because, he told his doctor, "I can't have this attached to me until I'm attached to someone else." But even after you're attached to someone else, you still don't want to appear dependent or weak. If a medical device improves your control but also makes you look debilitated or feel physically diminished, many patients will reject it.

As I get into bed one night with Sheryl, my pump, and my monitor, I remember a woman whose husband demanded that she remove her pump before sex, and she refused because, she said, "we're a threesome."

We're a foursome.

My continuous sensor is already a dinosaur. Two new systems that provide real-time glucose data (the Medtronic MiniMed Guardian RT and the DexCom STS) are now on the market, and a third (from Abbott Laboratories) is under FDA review. Medtronic has also introduced the first combined monitor and pump that not only displays real-time blood sugars but also recommends insulin doses.

All of the continuous monitors display glucose trend lines, which create a totally different framework for care. The numbers are no longer snapshots in time but part of a continuum; as one blogger wrote, "Watch the flow of the water, not the stones in the stream." In practice, patients would want to eat food while the glucose trend line is descending or would consider giving insulin when it's rising. In addition, the monitors sound an alarm when blood sugars go too low or too high, a blessing for diabetics who fear hypoglycemia while asleep.

Some doctors worry that continuous monitors will give patients too much information. They will be traumatized by the onslaught of numbers, and their physicians won't have time to sort through the data. The patients who have used the technology, mostly in studies, also warn that it isn't perfect: false alarms, inaccurate glucose readings, and calibration issues all create problems. But those are small sacrifices, they say, for what is essentially a window into your body.

The next step is the closed-loop system, in which sensor readings direct the pump to release insulin for normal blood sugars — a system that removes human error by removing individual decision-making. Recognizing its potential revolutionary impact, the Juvenile Diabetes Research Foundation decided in 2005 to spend $6.5 million for research on continuous sensors specifically and on an artificial pancreas generally.

The real questions are to what extent insurers will pay for these expensive systems and whether health care providers will be willing to train their patients how to use them. The most likely scenario? Insurers will resist because the technology will not be considered essential, and most doctors will conclude that they do not have the time or financial incentive to prescribe such a complicated device. In that case, continuous sensors will be a boon for the most sophisticated, engaged, and affluent patients, deepening the divide between the elite diabetics and the masses.

This evolution reminds me of television technology. In the old days, I plugged in the set, turned it on, and changed it to four or five channels — limited but simple. Now I have a satellite dish on my roof for a high-definition television with 799 channels that is connected to a receiver and a combined DVD and VCR player and requiring three different remotes. Red, white, and blue cables dangle from the back of the set like the arms of a rainbow octopus. When the system works, we receive a gorgeous picture with a wide selection of movies and sports. But all too often, we get SIGNAL NOT FOUND or DENIED ACCESS. Some problems are easily solved by simply resetting the receiver. Others can't be fixed — a storm, for example, that blocks the signal. Then there are times when I call the satellite help desk, and I'm told that one part of my TV isn't talking to another part. Why this is happening, she can't say, but sitting in Tulsa, she gets them talking again.

Technology will continue to improve the lives of diabetics, providing us with more rigorous therapies; and if an artificial pancreas is ever developed, I'll try it. But I suspect that before long, I'll be talking to someone in Tulsa.

At the end of the study, I'm given my glucose charts, and I have to admit that for all its frustrations, the sensor is remarkable. The charts reveal what I've never seen before: continual readings of my blood sugars, a glimpse beneath my skin. What I find is chaos. On the one hand, my overall glucose averages are consistent (week two: 138; week four: 153; week six: 132), and they produce A1c's in the low 6s. On the other hand, my scores gyrate wildly on any given day, reflecting my efforts to chase down high blood sugars with additional insulin. It's embarrassing, really. On April 3 at 6 A.M., I'm at 200. At 9:30 A.M., I'm at 305. By noon, I'm 50. On April 4, my blood sugars range from 400 to 70. April 18? I don't even want to talk about it. These days are extreme, but they underscore how even a vigilant patient with the best technology struggles to maintain control. I can make adjustments with my pump, but you can't fine-tune technology to achieve normalcy. You have to fine-tune your life, which means maintaining the same diet, insulin doses, exercise, emotions, and all the other variables that affect control. It's impossible, of course. At least the rapid insulin, the pump, and the meter allow me to make corrections quickly, but I think about the diabetics who don't have these tools or those patients in years past who were so poorly equipped to fight this battle. They never had a chance.

CHAPTER 10

Pushing Back the Horizon

IN 1972, GARY KLEIMAN was a freshman at Syracuse University when the haze began to cloud his vision. He first noticed it while driving to school from Florida, when fog seemed to drift around his car on the Pennsylvania Turnpike even though the sun was shining. Then he went to a basketball game and didn't know why the arena was so smoky and the players seemed to drift through clouds. While listening to music one night, he suddenly saw waves and ripples in the air, as if chiffon were floating in his dorm room. His album covers were out of focus. When he played basketball, his shots went awry. When he hit a lob playing tennis indoors, the ceiling seemed to be closing in on him.

Diabetic since he was six, Kleiman never suspected his illness was related to his sudden visual problems. He played on his college tennis team, a wiry blur beneath a thick head of curly brown hair, and he felt perfectly fine. Assuming he needed a new prescription for his glasses, he made an appointment with an ophthalmologist. After he struggled through the eye chart, the doctor entered the room, turned off the light, and picked up his ophthalmoscope. Looking in one eye and then the other, he saw blood vessels rampaging across the retina like weeds choking a garden.

"Son, you've got big troubles," he said.

Gary had diabetic retinopathy. His parents, summoned from Miami, took their son to a specialist at the University of Wisconsin ophthalmology clinic. When Gary swung open the clinic doors, he saw his own dark future. The corridor was lined with victims of the same disease, a tableau of eye patches, crutches, and misery that shocked the young man and his parents. An emaciated, legless man with black glasses sat in a wheelchair. A teenage boy sat holding a white-tipped cane. A woman in a beige pantsuit held her child, who was ca-ressing the mother's hair, cheeks, and eyes. The woman kept saying, "Isn't she beautiful?" She explained that she had lost her vision dur-ing pregnancy and had never seen her daughter.

Gary's ears felt hot, and his mother, Marge, saw a world of dark-ness in front of her and kept saying to herself, "This cannot be hap-pening."

When Gary saw the doctor, her bluntness bordered on cruelty as she told him his retinopathy was too advanced to be treated. "You'll be blind within a year and on dialysis in two" for kidney failure.

"How can you be sure?" Marge asked.

"Because that's the track record."

Gary told his parents that if he were ever put on a machine, they should pull the plug immediately and let him die. He was eighteen years old.

Kleiman is now fifty-three, having long ago lost all sight in his right eye and retaining partial vision in his left, his field of vision resembling a broken pencil surrounded by blackness. Two kidney transplants have kept him alive, though the immunosuppressant (or antirejection) drugs caused him to break bones in his feet seven times, to develop a cataract and cancerous lesions, and to acquire hepatitis C.

Kleiman has had better luck than most diabetics who develop se-rious complications at a young age. He is married, has two children, holds a good job, and has a full life. His experience shows how initia-tive, determination, and resilience — as well as extraordinary family support — allowed one person to overcome hardship and despair, denying the fate that his doctors had predicted. But his story also captures a much deeper theme in the history of this disease. The vic-tims of any illness want to be cured, but in diabetes, the push for the cure, at least in the last third of the twentieth century, was driven by parents, whose force shattered the complacency of the medical es-tablishment. Kleiman's parents were in the forefront of that move-ment — but so too was Kleiman himself. He was different. Most pa-tients relegate the cure to the back of their minds, a subject too daunting or depressing to ponder. But Kleiman didn't have that luxury. He sought the cure not to relieve him of his burdens but to save his own life. When he began his mission in the 1970s, a doctor told him that this generation could not be cured, but as Kleiman said, "If you're always planning for the next generation, you'll never get there."

In October 2002, he thought his dream had come true. Islet cell transplants had been the focus of a cure for thirty years, and now Kleiman received a transplant with a new cocktail of immunosuppressant drugs that had been hailed as a miracle breakthrough. After the procedure, his insulin function was restored. He could take the Metro rail without worrying about hypoglycemia, swim without having to remove an insulin pump, and just spend time with his family, relaxed and confident.

"I was unaware of the tension between testing or not testing, exercise or not exercise, this insulin or that insulin," he said. "It's an endless tug of war, with the strongest guys pulling the rope — and you're the rope! All of a sudden, they put the rope down, and the game is over."

But as Kleiman learned, the game had just begun.

Insulin's discovery prompted headlines about "a miracle cure," which not only misled the public but also hampered research. In 1956, Elliott Joslin underscored this impact in a remarkable statement. "There was almost a tinge of regret in several of us over the discovery of insulin," he wrote, "because it had come so soon." So soon? His own patients were dying. But in Joslin's view, scientists were making tremendous strides in diabetes research; after 1921, however, people told him, "Now, Dr. Joslin, you will have no more diabetics to treat." He feared the war against the disease was ending, comparing this "relaxation" to what he saw in France, on November 11, 1918, the day of the armistice. "In the twinkling of an eye, the whole atmosphere changed. Instantly salutes, which before 11 o'clock were alert and snappy, by noon had become informal and sloppy."

While Joslin acknowledged that his fears were overstated — research did continue — the belief that insulin was a suitable lifelong therapy held sway. This complacency was evident in the ADA's mission statement in 1940. Its goals included publishing a newsletter or magazine, lobbying legislators, writing press releases, organizing studies to determine the prevalence of diabetes, and encouraging the establishment of summer camps. What's striking is what's missing: finding a cure.

The ADA was founded by physicians, who should not have confused insulin with a cure. Why wasn't a cure one of the organization's original goals? A generous explanation is that doctors are trained to treat diseases, not cure them. A more cynical view holds that doctors profit from treatment, not cures. Regardless, the ADA became little more than a social club and referral service for physicians. At the same time, diabetes continued as a kind of underground disease, shrouded in myth and bereft of advocates. It would take a very different organization to bring the disease to public attention, to spotlight its ravages, to fight for a cure. It would take the parents of diabetic children.

When Lee Ducat's nine-year-old son, Larry, was diagnosed in 1969, she fainted. She knew that older people could have diabetes but had never heard of its affecting children. She went into a depression, wore dark glasses to hide her bloodshot eyes, and would hurry to Larry's school at recess to make sure he was okay. Larry's endocrinologist told them that insulin was the cure and that he would lead a long life, but Ducat sought out medical journals and discovered that diabetics could go blind or lose kidney function. "That stabbed me right in the heart," she recalled.

She wanted to do something. She lived near Philadelphia and visited its chapter of the ADA. Allowed to look through some files, she saw that all its members were doctors who treated adults and that the organization funded camps, but she didn't understand why money was not spent on research.

She wanted to talk to other parents of diabetic children, so she obtained names from Larry's doctor. She called five parents, but four out of the five wouldn't talk to her because they didn't want anyone to know their children had the disease. She was further troubled when she took Larry for a checkup and told the doctor that she had been reading about complications. "We want to keep that a secret," the doctor said, "because we don't want to upset the children."

How can you cure a disease, Ducat wondered, if nobody knows it exists?

Ducat, who was thirty-seven, had had experience as a fundraiser, and she asked the doctor if more research could be undertaken if he had more money. He said yes. "If money is what you need," she told him, "then I'll get you money."

She asked her husband's lawyers to register a nonprofit organization with the State of Pennsylvania. She had a cocktail party and, through patient lists and word of mouth, attracted representatives from about sixty families. Membership cards were distributed; dues were collected. This new volunteer organization called itself the Juvenile Diabetes Foundation (JDF), and its members discussed their goal: to find a cure.

Ducat, who was also a professional singer and performer, recruited a publicity firm to work pro bono — one of its executives had a diabetic child — and her efforts were soon described in newspapers and magazines. Children had diabetes. They needed daily injections. They could develop serious complications. Who knew? The publicity generated calls from parents in other cities, typically angry, heartbroken mothers determined to challenge the indifference of health care providers, speak the truth about the disease, and cure the damn thing. They were the original desperate housewives.

Sometimes they were met with resistance. In New York, several parents tried to join the ADA, which invited them to a meeting but made them wait in a file room. "The doctors looked at us and said, 'What do you know? You're parents,'" recalled Carol Lurie, whose son was diagnosed when he was ten. They heard about the JDF, and Lurie's husband, Erwin, attended a meeting. "They've got spunk and spirit," he reported.

The New York chapter was soon born, along with chapters in New Jersey, Miami, and Washington, D.C. All the spunk and spirit in the world, however, couldn't move the medical and political establishments to respond to its demands for more research. After all, the JDF's constituents — diabetic children — lacked power, stature, and numbers. But the parents knew how to get attention. As one of them said, "We will follow the golden rule. He who has the gold makes the rules."

The JDF used as its model the largest and arguably most successful voluntary health organization of all time, the National Foundation for Infantile Paralysis. Founded in 1938 by President Roosevelt to raise funds for curing polio, it used celebrities to promote the cause, including Eddie Cantor, Jack Benny, Bing Crosby, Rudy Vallee, even the Lone Ranger. Its most effective advocate was Roosevelt himself, whose status as a polio survivor allowed the organization to hatch its March of Dimes campaign, urging Americans "to send their dimes directly to the President at the White House . . . to show our president we are with him in this battle." While Roosevelt personified the disease, donors were also moved by newspaper photographs of suffering children, who were struggling in leg braces, sitting in wheelchairs, or encased in iron lungs. The foundation used some of these images for its own fundraising while also highlighting — if not actually exaggerating — the scope of the disease. At the same time, the group assumed donors wouldn't send money to a lost cause, so it produced pamphlets and articles emphasizing the progress researchers were making in finding a cure. In doing so, according to David M. Oshinsky, it "skillfully mixed the public's dread of polio into its larger message of inevitable triumph."

Movie stars, heartbreak, and victory: if it worked for polio, it might work for diabetes as well. The JDF's chapters consolidated, created a national headquarters in New York, and eventually changed its name to the Juvenile Diabetes Research Foundation (JDRF), enlisting scientific advisers and developing educational programs and support groups. Its fundraising success allowed it to evolve from a volunteer organization to a professional bureaucracy, with sophisticated marketing and publicity efforts. It recruited celebrities like the glamorous socialite Dina Merrill (her son had diabetes) and Mary Tyler Moore (she has type 1). It featured children with needles or in wheelchairs, raising the dark specter of complications while insisting that researchers were homing in on a cure. It lobbied lawmakers in Washington, with Ducat and Lurie visiting senators, congressmen, and presidential aides, spurring increased funding by the government. Over the years, it sent children to Capitol Hill to describe the horrors of diabetes while holding galas, rallies, and marches. But its focus never changed. It was about the money.

Not everyone approved of these tactics. Some patients believed that the emphasis on complications was unduly negative and the use of children was sentimental exploitation. The organization, according to this view, reveled in misery. Michael Bliss said the JDRF never wanted to hear his lecture on insulin's discovery because it was too "positive." Others would later object to the group's sunny projections about impending cures that never materialized. But the JDRF succeeded in bringing diabetes out of the closet, and its clarity of purpose turned it into a financial juggernaut. It raised $900 million in thirty-five years and announced in fiscal year 2005 a campaign to raise $1 billion in five years. It also forced the ADA to broaden its outlook, funding outreach programs, advocacy efforts, and researchers. The ADA's current motto reflects its changed priorities: "Cure. Care. Commitment."

The early years of the JDF coincided with researchers' fledgling efforts to eliminate diabetes through islet transplantation, an effort that occupied some of the leading diabetes scientists for decades to come.

The idea of transplantation wasn't new. In 1893, when the sugar-destroying properties of the pancreas were still under debate, an English surgeon tried to transplant pancreatic fragments from a sheep into a fifteen-year-old diabetic boy. The first pancreas was transplanted in 1966 at the University of Minnesota; the patient's blood sugar fell, but she died three months later from a pulmonary embolism. The safety of pancreas transplants has improved, of course, and has always remained an option, primarily for diabetics who are also receiving a kidney transplant and are already on immunosuppressant drugs. But whole-organ transplants at best were highly impractical in curing diabetes. Only 2 percent of the organ consists of islet cells, which account for the insulin-producing beta cells. The rest of the pancreas consists of tissue that secretes powerful digestive enzymes. It didn't make sense to remove an entire organ when only 2 percent was damaged. The surgery itself was complicated and expensive, and postoperative complications sometimes affected the other 98 percent, the healthy part, of the pancreas.

Paul Lacy wasn't thinking about curing diabetes when, in the late 1960s, he tried to extract islet cells from a rat. Lacy, the head of the pathology department at Washington University in St. Louis, was trying to understand the mechanics of hormone secretion from beta cells, but once he collected the islets from the rats, he couldn't resist seeing if they could control blood sugar levels. In 1972, he successfully transplanted the cells into a diabetic rat, returning its blood sugar to normal and suddenly raising hopes that a cure was possible. It seemed to make sense: if your islets don't work, get new islets; and an islet transplant would be far more elegant than using an entire pancreas.

Lacy himself was an avuncular figure who conveyed a genuine sensitivity for the plight of diabetics, recognizing that insulin was a crude therapy and that complications destroyed lives. He also brimmed with confidence, telling parents — including my own (we lived nearby) — that a cure was five years away. His optimism, added to that of other researchers in the field, buoyed the nascent JDF. Lacy himself became good friends with Lee Ducat, and in the years following America's dramatic success in space — when all things seemed possible — islet transplantation became the moon shot of the diabetes community.

But Lacy was naïve about the impediments to using his experiment with human subjects. To begin with, the way he isolated islet cells in rats didn't work on comparable human cells. He used scissors, blenders, knives, and meat grinders; nothing worked until an isolation technique was developed in the mid-1980s. The first human trial began in 1985.

The islets were not placed in the remote pancreas but were delivered by making a small incision near the navel and feeding them through a tube into the network of blood vessels extending from the portal vein of the liver. The injected cells were to form small nests in the liver and function like the cells of a healthy pancreas, sensing from moment to moment the body's glucose level and releasing the correct amount of insulin.

The grafts, however, didn't work. In the coming years, more than four hundred patients received transplants, but they all failed to maintain normal glycemia. Many suspected that the immunosuppressant drugs, which included steroids, were also toxic to the islet cells. Some scientists concluded that the transplants were futile. With only 5,000 organ donors a year, it was hard to find a cadaveric pancreas whose islets were compatible with those of a test subject, let alone to find cells that actually worked. In 1996, the National Institutes of Health, the primary financial supporter of biomedical research, devoted less than 3 percent of its diabetes budget to research on islet cell transplants.

In 1995 Lacy, recently retired, wrote that "islet transplantation is conceptually simple but difficult to implement," noting that "islets in most patients have been unable to control blood sugar levels completely or have lost some of their activity after three years or less." But he remained as bullish as ever, perhaps due to recent improvements in immunosuppressant drugs: "There is good reason to think this potentially curative therapy will be available to many patients within the next five years."

Many teenagers go through periods of rebellion. For diabetic teenagers, the tools of their rebellion are perilously handy: they can stop taking their insulin or ignore their diet.

Gary Kleiman drank Cokes while wrestling with the contradictions of his disease: he had an invisible disorder, but the doctors said he could live a normal life. If true, then the problem itself must not be that serious. So he played tennis for five straight hours, ate rich desserts, and refused to test his urine. He also exploited his condition, blithely missing classes, demanding unnecessary snacks, getting longer rest periods, and shirking responsibilities. "I got away with murder — and suicide," he later said.

In college, a severe flu put him in the hospital, where doctors found a trace of albumin in his urine, the first sign of kidney disease. It should have put them on high alert. Instead, they told Gary not to worry. Two months later, his retinopathy was diagnosed. He received laser treatments — still new and imprecise — to destroy the bad eye vessels before they consumed the healthy ones. On one occasion, his head was strapped to a machine while drops were administered. Then an object like a jeweler's loupe was poked into his eye and a laser beam shot into the center of his head. He could feel the beam ticking and hear its *rat-tat-tat*. It sounded like an automatic stapler.

Gary dropped out of college, lost his driver's license, and severed friendships. His blood pressure was high, his kidney problems were progressing, and his diabetes was still out of control. Desperate, his parents heard that diabetics could reverse their complications with a special rice diet — not unlike the special "oat cure" in the nineteenth century. Gary checked into the "Rice House" at Duke University, often used by corpulent celebrities to lose weight. But his health didn't improve; after nine months in which his only highlight was sharing some rice with Buddy Hackett, Gary returned to Miami.

In 1972, a family friend was made the executor of an estate that gave him almost $25,000 to donate to a worthy cause. Aware of Gary's condition, he offered to contribute the money to diabetes, but neither he nor the Kleimans knew where to send it. Gary's parents soon heard about the new Juvenile Diabetes Foundation. Joining it, they met Daniel Mintz, a researcher at the University of Miami, who was working on islet transplantation. He convinced them that islets offered the best hope for a cure, and his personal humility — a scientist who spoke to parents, not at them, and who acknowledged that he, not being diabetic, could never truly understand the burden of the disease — inspired trust. The Kleimans gave him the $25,000. After the JDF became a national organization, the South Florida chapter withdrew so it could support Mintz's work exclusively. As Gary's father, Marty, told his wife, "I'm going to devote every minute that I have to support this man's research until a cure is found or until I die, whichever comes first."

Like other JDF parents around the country, the Kleimans raised money by drawing attention to the disease. Marge began giving speeches in South Florida, her husband standing by her side. She never used notes and never asked for money directly. Rather, she told the story of her own family and others affected by the disease.

"Imagine that you have a nine-month-old child, and you don't know if he's sleeping because he's tired or because he has too much insulin," she would say. "There is no way for a parent to know and you can't ask the child; and to test his urine, you have to take off his wet diaper, wring out a few drops, and figure it out."

When the organization placed a full-page newspaper ad to publicize its cause, she discovered how deeply in denial were the leading lights of the diabetes community. The ad itself featured a little boy in a crib holding a glass syringe, and it described the complications that could arise, including kidney failure, heart disease, blindness, and amputation. The headline read: THE QUIET KILLER.

Marge Kleiman was working in the JDF office when the ad appeared, and the telephone rang.

"I'm Charles Best," the caller said, "and I discovered insulin." Now retired, he had become an iconic figure who, after Fred Banting died in 1941, carried the mantle for the team that had discovered insulin. He had been praised by the pope, the queen of England, and other heads of state, while reveling in his honors from grateful diabetics, medical organizations, and universities. He gave the keynote address at the ADA's first meeting, later served as its president, and was its honorary president until he died in 1978. He happened to be in Miami that day, and he was outraged by the newspaper ad that was tarnishing the victorious narrative of diabetes.

"What kind of propaganda are you using?" he screamed. "You're frightening people! This is not the way it is!" Best read each complication named, then denied they existed. "You're telling lies!"

But Kleiman knew better. "Dr. Best, what you did was wonderful," she said. "It allowed people to live longer. But we're not trying to frighten people. If you tell the truth, maybe they can avoid these complications. Please don't tell us to keep quiet."

Gary also became a highly visible spokesman; he was featured in news stories, appeared in public service announcements, and gave speeches. Articulate and smart, he was a powerful symbol of diabetic devastation. In 1973, he appeared before Congress when it was considering an appropriation for diabetes research. "Gentlemen," he began,

> You'll have to excuse my not having a prepared statement, but if I had one, I couldn't read it. I am going blind from diabetes. When I was six years old, I became one of the statistics you've been hearing about. A child with diabetes. Then at eighteen, I was another statistic, a young man with complications from diabetes. I now can become another statistic. There is no choice. I can be one of the stricken who will be dead ten years from now or I can be the best statistic of all — a person cured of diabetes. By funding the necessary research, you can make that possible.

His struggles had also made him depressed, and his odd celebrity did little to ease that problem. His only goal was to survive the year, and when he did that, to survive the next year. "I'm dying piece by piece," he told a friend. He was eventually hired to assist patients at the University of Miami's Diabetes Research Institute (DRI), which had been created in part by the funds his parents raised and had become a permanent home for Mintz's work. Kleiman related well to children. No one knew how to mismanage their disease better than he, and the price he had paid was clear. As Jay Skyler, a professor of medicine at the university, said, "He was a good role model, in spite of himself."

In 1982 his kidneys shut down, but he avoided dialysis by receiving a kidney from his mother. Before she went under the anesthesia, Marge told the doctor, "You know, having a baby is easy. It's the maintenance that's tough." When Gary awoke, he wrote:

Dear Mom,
    I want to thank you for your generous gift. It fits, it works, and honestly, it was exactly what I needed.
Love, Gary

The breakthrough everyone was hoping for came from the same country, Canada, where insulin had been discovered.

Scientists at the University of Alberta in Edmonton noted that islet cell transplants had a much higher success rate among patients who suffered from chronic inflammation of the pancreas but did not have diabetes. In these cases, the procedure involved removing the patient's pancreas, extracting a large number of islets, then infusing the cells back into the patient. No immunosuppressive medication was necessary — the patient was receiving his own cells. This seemed to confirm what many had already suspected: the antirejection drugs, particularly cyclosporine and corticosteroids, were lethal to transplanted islets.

The researchers began working on a new cocktail of drugs using rapamycin, which had recently begun being tested on humans and appeared to work differently from other comparable agents. They also believed that transplants were failing because patients were not receiving enough high-quality islets. Scientists at the DRI in Miami, however, had recently developed a mixture of enzymes that broke down the tissue of the pancreas, freeing the embedded islets without injuring them. This technique was used by the Canadians, along with a new mix of immunosuppressants, to create the "Edmonton protocol."

In the summer of 2000, they reported that they'd given islets to seven diabetics under the new regimen, and after roughly a year, all were still off insulin — by far the best success rate ever. It quickly became the stuff of medical legend, affirming three decades of effort to turn islets into a cure. "Here was this absolute miracle," said David M. Nathan, director of the diabetes center at Mass. General Hospital.

The JDRF declared islets a top priority, and over the next four years poured $225 million into the field, with commitments for more. The organization unabashedly called the Edmonton protocol a success, using it to attract additional contributions. Hospitals in the United States and Europe raced to set up islet transplant centers. Journalists wrote or broadcast stories about diabetics who no longer had to take shots. One woman, Ellen Berty, wrote a book: *I Used to Have Type 1 Diabetes: Kiss My Islets.*

By the time the Edmonton protocol was announced, Gary Kleiman had improved his control with an insulin pump, but other changes in his life confirmed his desire for islet cells.

He had expanded his part-time counseling job at the diabetes institute, becoming its executive director of medical development. As his life became stable, he was less upset about his impaired vision than his thinning hairline. He created a manageable if insular existence, living alone in Miami, within walking distance of most of the things he needed. The institute gave him a car and a driver to get to and from work. His secretary read him his mail.

Kleiman thought little about the future until he met Chris McAliley in 1989. She was a tall, attractive lawyer — she later became a judge — and an improbable romance blossomed.

"You don't want to be with someone who's blind and has a kidney problem," he told her.

"That's not the problem," she said, "but you're shorter than me and going bald."

Some of her friends thought she was crazy, including a physician who wanted to set her up with a cardiologist. The friend told her that Gary "would die young, would never be able to have children, and even if you did want children, that would be irresponsible."

But Gary gave Chris something she had never had — an extended family — and his parents' faith in a cure buoyed her own spirits. The family was a team that worked together and never gave up. The idea touched Chris, who figured she could always marry "a healthy specimen of a man" who could get hit by a bus and not know how to deal with it. Gary had proven he could overcome almost any trauma.

Gary had learned that the immunosuppressant drugs from his kidney transplant, which included steroids, had probably made him sterile, so the couple resigned themselves to not having their own children. But when they were planning their wedding in 1992, Chris discovered she was pregnant.

"I thought you were sterile," she said.

"Doctors don't know everything," Gary quipped.

They married, had two (nondiabetic) sons, bought a house in the suburbs with a swimming pool, and got a dog. They juggled day care, vacations, and carpools, and Gary for the first time bought life insurance and began a retirement plan. After years of relative isolation, he entered the mainstream, but the more "normal" he tried to make his life, the more deficient he felt. Unlike other fathers, he couldn't pick up the groceries for dinner, play catch with his sons, take them to ball games, or help with homework. He couldn't jump into the pool without removing his pump. He tripped over toys.

Worst of all, he feared for the safety of his youngsters. If he were with them on a busy street or crowded park and his blood sugar dropped too low, they would be in jeopardy. His new responsibilities also made him more nervous about the risks to himself. Gary was always second-guessing himself. Sometimes he would feel wobbly, but he wasn't sure if that was from his kidney medication or low blood sugar. "I just didn't know how I was feeling," he said. These fears caused him to walk around the mall or even his own office with a can of Coke, constantly sipping to preempt hypoglycemic incidents. But he couldn't eliminate all of them. One late-night reaction forced Chris to call paramedics.

Gary knew the solution: restoring his body's insulin function without hypoglycemia — in short, islet cells. In 1999, when the early results of the Edmonton protocol were released, Gary immediately lined up for the experimental trials in this country. But he had one more crisis to face.

His new kidney began to fail, and his health deteriorated. He was losing weight and felt fatigued; some days he couldn't get out of bed. He had been using the kidney for nineteen years, far longer than the expected five, and he needed another one to avoid dialysis or life-threatening complications.

Finding a matching cadaveric kidney is difficult, given the paucity of donors, and risks of rejection are high. A kidney from a living person is more likely to be accepted, particularly from a blood relative. But Gary's mother could not part with her second kidney, and his father had died. That left one last living relative — his brother, Glenn, a chief executive of a communications company in Milwaukee. He agreed to donate a kidney: the team had come through one more time.

Before the transplant, Gary requested that he be placed on the same immunosuppressants used in the Edmonton protocol. If his body successfully adjusted to the drugs, it would be ready to accept the cells when they became available.

In October 2002, Kleiman got word that a police officer had been killed in a motorcycle accident; the donated pancreas matched his tissue type. The officer had been a big man, which meant his pancreas had an unusually large number of healthy islets — a good omen for a transplant. On November 1, Gary lay shirtless on the operating table in a special radiology suite at the University of Miami, surrounded by beeping monitors, X-ray machines, glucose monitors, and a university film crew. Doctors, medics, nurses, and students, all wearing gowns and caps, packed the room; Gary's wife and mother peered through a window. One person who couldn't watch was Dan Mintz, who had more than a researcher's interest in the procedure. After Gary's father died, he married Gary's mother, and now his love for his stepson gripped him with fear.

Gary received light sedation and local anesthesia, then a yellow antibacteria ointment was spread on his stomach and a catheter was inserted beneath his ribs. A nurse carried in small thermos coolers and withdrew a plastic bag filled with yellowish liquid and faint dots — the islet cells "looked like little ants running around," his mother later said. Kleiman was awake as the bag was attached to an IV tube and the islets dripped into his liver. His wife and mother cried; so too did some of the medical staff, who had known him for almost thirty years. Gary thought about his father, about his many diabetic friends who hadn't made it, and about the doctor who cautioned his parents when they joined the JDF: it's too late for this generation.

But it was not too late for Gary Kleiman. The transplant took twenty minutes, with his cracking at the end: "They had to build an entire building for this?" Everyone laughed.

When he got home, Chris jumped into their pool, gazed at the sky, and thought, "The horizon has just been pushed back" — an echo of Elliott Joslin's comment when insulin was discovered: "The promised land is in plain view."

But the reality of islet cell transplants proved more complicated. At some centers, the results were initially good, with 80 to 90 percent of patients independent of insulin a year after their transplant. At some point, however, the cells begin to erode — what James Shapiro, the lead investigator in Edmonton, describes as an "inexorable" loss of insulin function. In 2005, he said that of his eighty-two patients who had received the transplant, 70 to 80 percent had achieved normal glycemia immediately after the procedure, but that number dropped to 11 percent after five years. Everyone else returned to insulin injections.

What happened? It's possible that the immunosuppressant drugs, while less toxic than previous ones, still harm the islets. The underlying immune system that killed the original islets may also be destroying the transplanted ones. Patients may also be receiving too few islets. A normal pancreas has roughly 1 million cells, but current techniques only allow about 400,000 to be extracted from a donor pancreas, and unknown numbers die soon after they're transplanted.

The procedure has other problems. Pancreata are scarce and costly. Fewer than 2,000 are donated each year; most go to whole-organ pancreas transplants and cost from $15,000 to $20,000 apiece. The islet cell transplant itself, meanwhile, costs up to $200,000 for one patient in the first year, and the necessary drugs add another $30,000 a year.*

While the loss of islet function is disappointing, the Edmonton protocol's side effects have been far more alarming. In 2004, the NIH's Collaborative Islet Transplant Registry reported that of the eighty-six islet recipients it surveyed, it had catalogued twenty serious adverse events linked to the transplant. They include four cases of life-threatening neutropenia, a depletion of white blood cells caused by antirejection drugs.

In 2004, the Edmonton team released the data on its first forty-five transplant recipients. Of the five patients followed for four years, two had "quite bad renal outcomes," including one who had to go on dialysis. Overall, a third of the forty-five had high levels of protein in their urine, which is normally a sign of declining kidney function. Some of those patients may have developed diabetes-related kidney problems anyway, but the number was still high.

In 2000, the NIH's National Institute of Diabetes and Digestive and Kidney Diseases had received more than $1 million to launch a transplant program using the Edmonton protocol. From December 2000 to June 2001, six women received transplants, and the initial results were encouraging. Four of the six became independent of insulin, and three stayed that way for at least eighteen months; those who needed insulin no longer suffered from hypoglycemia. This phenomenon was seen in other studies: even when the islets didn't restore complete glycemic control, they allowed patients to achieve better control with less exogenous insulin (from outside the body) and fewer lows.

But two patients in the NIH study, including one no longer using insulin, had to discontinue the immunosuppressants because of intolerable side effects, including deteriorating kidney function. Ellen Berty, who wrote the book about her successful transplant, had mouth ulcers so large that dentists at NIH photographed them for a

---

* Because the procedure is still considered experimental, it is funded by the NIH, the JDRF, the DRI Foundation, and some pharmaceutical companies.

textbook. Fearing that the transplants were doing more harm than good, the investigators ended the trial early.

By the end of 2005, 550 patients had received transplants from fifty institutions worldwide, but the procedure was clearly not practical for most diabetics. To qualify for a transplant in Edmonton, for example, patients already had to be on immunosuppressants. If not, they had to have had several bad lows, hypoglycemic unawareness, or unusually volatile blood sugars. Of the 1,700 who had been screened for transplants, only 6 percent met the criteria. "Every patient we take on, they're near death's door or in desperate straits," Shapiro said.

Therapeutic advances in pumps and continuous glucose sensors have also altered the risk-reward ratio in transplants. Liver, heart, and kidney transplants can be life-saving operations, making any side effects of the immunosuppressants an unfortunate but necessary tradeoff. For most diabetics, islet cell transplants would not save their life. At best, they can stabilize blood sugars and reduce lows, which is no small achievement. But they would also subject a patient to a lifetime of immunosuppression — a "cure" that could be worse than the disease.

Because Gary Kleiman was already taking the immunosuppressant drugs for his kidney transplant, he avoided the rejection problem that had thwarted other patients. Three years after his transplant, he considers it a success. His A1c's used to be between 7.6 and 7.8; now most are in the 5's — in other words, blood sugars near those of a nondiabetic. Equally important, he no longer experiences extreme highs and lows, which also reduces the risk of further complications. The elimination of hypoglycemia is a huge benefit. In the first year after the transplant, when the islets were at their strongest, he even went to an Italian restaurant and ate pasta and tiramisù. His blood sugar didn't spike.

Kleiman has been featured prominently in newspapers, magazines, and on television, which also reflects well on his employer. His advocacy of islet cells is a reprise, of sorts, of his role as an early spokesperson for the South Florida JDF, when he represented the damaging impact of the disease. Now he is considered a success story.

But Kleiman knows that he's not cured, recognizing that the islets are fragile and can ultimately perish. A year after the transplant, he was in the back seat of a taxi that crashed into another car, slamming him against the front seat, cutting his lip, and shaking him up. The accident traumatized the islets, forcing him to take fast-acting insulin. While that need soon ended, he decided to take a daily injection of Lantus — long-lasting "background" insulin — to help preserve the islets. For someone who took shots for more than forty years, the sacrifice was inconsequential, but it was an obvious reminder that he still has the disease. He is now considering a "fill-up" — another injection of islets before his current ones die completely.

Dan Mintz describes the outcome as short of a cure but a "satisfying end point," and Kleiman concurs. "I don't view this as a cure at all," he says. "I view it as an imperfect but amazing therapy."

Given his bleak prognosis at eighteen and his myriad health problems since, Kleiman attributes his survival to the exceptional advances in medical research. But he also acknowledges the unfulfilled promises of those early days, when his family committed itself to defeating the disease.

"When we started this way back we were very naïve, but it was a very naïve time," he says. "You had these researchers who were really believing if you can do it in mice, you can do it in humans. Then, as things progress, you get frustrated, and then you get more realistic. And you realize that every time you open a door, there are two more doors."

CHAPTER 11

The Magical Beta Cell

AS ONE OF AMERICA'S leading embryonic biologists, Douglas A. Melton spent his time investigating frogs, allowing him, as he says, to understand "the basic biology of how life develops." With a tenured position at Harvard University and funding from the prestigious Howard Hughes Medical Institute, Melton had the luxury of doing research with little practical value. "It was a selfish thing," Melton says, "just as a musician is interested in music or an artist is interested in art. They do it for themselves."

Melton's self-absorption ended in 1991, shortly after the birth of his son, Sam. Melton and his wife, Gail O'Keefe, had noticed that Sam was less active and "smiley" than their first child had been, and they were concerned about a stubborn diaper rash. Doug blamed the cloth diapers that Gail used — he favored disposables. They tried creams and changing Sam more often, but the rash persisted.

Early one morning, Sam woke up vomiting and panting. Doug thought it was a flu and went to work, but Gail knew something was terribly wrong. She took the baby to a pediatrician near Doug's office; a nurse couldn't diagnose the problem but said it was more than a flu. A few hours later, Doug and Gail were in Children's Hospital in Boston; Sam, lying on a small metal table, continued to breathe heavily. He was completely dehydrated and had begun to slip into unconsciousness. The doctors ran some tests and did a barium enema, but they had no idea what the problem was. When they finally asked Doug and Gail to leave the room, Gail asked if her child was going to survive.

"Honey," a doctor said, "I don't think so."

Then a miracle happened: Sam urinated. An alert nurse grabbed a test strip, dipped it in the puddle, and discovered the fluid was soaked with sugar — the very sugar that had been causing the rash. The doctors scrambled to revive the baby; at six months, he was the youngest child diagnosed with diabetes in the hospital's venerable history.

A doctor told Doug and Gail the news: "He's going to need to take insulin shots."

"For how long?" Doug asked.

Overnight, Melton's intense but abstract quest for knowledge changed to a parent's primal urge to cure his child. When he returned to his lab, he gave an emotional talk to his team. He said he would halt all of their current research and redirect it toward type 1 diabetes. Reading the literature, he concluded such a cure had to focus on two areas: the immune system and the beta cell. He didn't believe he could contribute to autoimmunity — he says there wasn't then, and isn't today, a good "theoretical construct" for that problem — but he could work on the beta cell, what he calls "the supply problem." Melton wanted to create beta cells that wouldn't trigger an immune response the way transplanted islets did.

While Melton believes the beta cell should be a source of endless inquiry — "How can it so exquisitely measure sugar and secrete just the right amount of insulin?" — he says that diabetes research has lagged because the disease itself is viewed as "half-solved" with insulin. The brightest scientists work on cancer or neurobiology instead. He wants to recruit young investigators by presenting diabetes as both a serious disease and an intellectually interesting problem. "If cells can turn on themselves — autoimmunity — why do we have so few autoimmune diseases?" he asks. "Right next to the beta cell is the alpha cell, which makes glucagon. Why don't we have other people who have autoimmunity to that? Or autoimmunity against our heart muscle or our brain? What is so weird about the beta cell? That's a really good puzzle."

Like most scientists, Melton would like to divine these mysteries in the seclusion of his lab, but the deeper he drilled into the molecular underpinnings of the body, the closer he probed the origins of life. Perhaps it was inevitable that his work would run headlong into controversy, thrusting an obscure frog biologist into the crosshairs of a bitter national debate about science, religion, and morality.

Stem cells form the literal foundation of human life, emerging just days after an egg is fertilized with the unique ability to transform into a wide variety of tissue, from bone to brain to liver. Stem cells fall into two categories: adult stem cells are present in the body throughout one's life, centered in areas like the spleen, bone marrow, and skeletal muscle, where they create new cells to replace those lost to the daily wear-and-tear of living. So-called embryonic stem cells are far more versatile. They form in the embryo about four days after fertilization and can become virtually any cell in the body. Unlike adult stem cells, however, they completely disappear well before birth.

Scientists have known about stem cells for more than a hundred years, but only in 1998 were human embryonic stem cells first isolated in a laboratory and subject to direct investigation. Their curative and therapeutic potential is enormous, because their ability to divide, as well as their versatility, offers a renewable source of replacement cells. In theory, they could be used to create dopamine-producing cells, the absence of which causes Parkinson's disease, new cardiac tissue for heart attack victims, new skin tissue for burn victims, or new blood and marrow cells for those who have undergone chemotherapy. For diabetics, they could be used to create beta cells as well as new heart muscle, kidneys, or other organs damaged by the disease. They could even play a role in type 2 diabetes if investigators can determine how to guide or manipulate fat cells so that patients can lose weight or become more sensitive to insulin.

But embryonic stem cell research deeply offends those who believe that a fertilized egg constitutes life, its use for science therefore tantamount to murder. It doesn't matter, to critics, that the actual eggs used by investigators are surplus embryos from fertility clinics, which would otherwise be discarded. The opposition to this research, particularly among religious conservatives, carries symbolic weight: if destroying a fertilized egg, at any stage for whatever reason, were permitted, it would undermine their argument against abortion that life begins at conception.

Proponents of the research say that embryonic stem cells are microscopic dots that "have fewer human characteristics than a potato," and protecting them — instead of seeking to cure diseases — is lunacy. While the cells could relieve many diseases (spinal cord injuries, lymphoma, multiple sclerosis, Alzheimer's), diabetic children are often highlighted as the most sympathetic victims. "I know a child," said Ron Reagan, the president's son, to the Democratic National Convention in 2004. "She has fingers and toes. She has a mind. She has memories. She has hopes. She has juvenile diabetes." He said she wore an insulin pump with rhinestones and had learned to sleep through blood draws in the wee hours of the morning. "She's very brave . . . [but] understands full well the progress of her disease and what that might ultimately mean: blindness, amputation, diabetic coma. Every day, she fights to have a future."

Doug Melton doesn't look like a crusader. A lean figure with wire-rim glasses, graying hair, and a gentle voice, he projects middle-age decorum, a scientist who leaves nothing to chance. When he opens the door of the restroom at the Harvard Faculty Club, he wraps his hand in his scarf so his skin never touches the knob, which may have germs. His understated persona, however, belies a fierce intellect and a supreme self-confidence, which lead to biting criticism of his peers. He ridicules scientists, for example, who believe that adult stem cells can be found in the pancreas — "It's a waste of precious time and effort." He says the research on diabetic complications is "pathetically bad," with investigators "spending huge amounts of money on things that are really just a shot in the dark." He describes experiments indicating that beta cells can regenerate "not just flawed, but laughably flawed." As Gail says about her husband, who also has an advanced degree in philosophy, "He doesn't like to not know something, and he doesn't like to be wrong."

Sam's diagnosis plunged the family into despair and uncertainty. Gail went through what she calls her "black winter," struggling with blood checks and insulin doses, unclear about his diet, petrified when he got sick — he once vomited on an airplane (vomiting can lead to hypoglycemia) — but determined to maintain a physical and emotional connection. (She nursed him until he was three and a half.)

Nonetheless, she says, "I was always worried, and I felt as though I had been robbed of having an infant."

Gail also resented Doug's lack of direct involvement in Sam's care. She was in graduate school at the time, studying environmental policy, but that came to a halt. "I didn't have time for anything else, but Doug could walk out in the morning and leave for nine or ten hours or go on a business trip," she says. "I was livid. I grew up in this era when marriage should be a partnership, and I wanted some sharing of this enormous load, but he wasn't good about getting up and doing the blood test and figuring out the insulin."

While she grieved for Sam, her husband was emotionally detached. "Doug could compartmentalize and not really relate to it," she says. "If he hadn't let diabetes into any of his compartments, I don't think I would have stayed with him." His response was to throw himself into his research — his way of sharing the burden. "He was at the top of his game in developmental molecular biology," Gail says, "and now he was starting from scratch and was learning something that he didn't feel he had a chance to succeed in. It was very courageous of him."

Melton acknowledges "a strain in my family and a tension between my wife and me . . . She gets frustrated with me." He also recognizes his wife's contributions. "Gail is really Sam's pancreas," he says. When Sam is out of the house, he text-messages his blood sugar numbers to her, and Gail responds with the insulin dose. The rapport is so close, Doug says, that Gail has a "sixth sense" about her son. "She'll wake up at four in the morning and say, 'I think Sam is low.' And sure enough, she's right. She noticed something that neither Sam nor I noticed, and it just kind of jells in her mind in the middle of the night."

Melton began his diabetes research by examining embryonic stem cells in frogs, chickens, zebra fish, and mice; then, when the University of Wisconsin isolated human embryonic stem cells, he asked to work on them. The university, however, demanded that he adhere to commercial and scientific restrictions that he felt would inhibit his research. Instead, he found that he could import the cells from scientists in Israel.

Melton believes the embryonic stem cells provide the best way to glimpse human development, comparing his approach to his arriving at the scene of a car accident. He has no idea why the car crashed, and he might be able to patch it together again, at least temporarily. But he would rather take the car to the factory and understand how it was made. Only then could he recognize fundamental design flaws. In effect, he wants to take the beta cell back to the factory. "So much of the medical community has studied defective beta cells," he says. "I want to focus on how to make a healthy one."

But that challenge entails nothing less than studying the mystery of life. Every person begins as a single, fertilized egg, which, after implanting in the walls of the uterus, develops three layers of cells. The cells from one layer, the endoderm, will become the lungs, the liver, and the pancreas. The cells then specialize further, with only a small number becoming beta cells. Melton compares this process to a kindergartener who might grow up to become a writer instead of a lawyer. What influenced that student to become one and not the other? You don't have to identify every event that shaped the child's life, but you need enough information so you can direct other students to choose that same path.

Similarly, Melton doesn't need to understand every event that causes an embryonic stem cell to become a beta cell instead of a bone cell or a blood cell, but he needs to identify the few definitive genetic signals that produce that outcome. That could make it possible to create beta cells for transplantation, using embryonic stem cells as a virtually unlimited source and thereby negating the supply problem that surgeons currently have with islets or pancreata.

The actual research, however, has been painfully slow, because neither Melton nor anyone else knows how to coax a beta cell from an embryonic cell. When Melton, collaborating with the Israeli scientists, tested the effects of various growth factors, or proteins, on the cells, he found that these chemical signals could nudge the cells in one direction or another, but his control was limited. The cells, according to Melton, behaved like "popcorn," morphing unpredictably into various cell types.

His frustration, however, did not diminish his passion for the research, and his advocacy soon put him in the public eye. In 1999, when the federal government considered what limits, if any, it

should impose on embryonic stem cell research, Melton testified before the U.S. Senate, speaking first as a parent. When Sam is asleep,
he said, "we wonder, is his blood sugar too low? Will he find the middle ground between a 'low,' or coma, and being too 'high' in the
morning? I can't recall a single night since Sam was diagnosed that
we slept peacefully, free of the worry that the balance between food,
insulin, and exercise was not good enough. I'm unwilling to accept
the enormity of the medical and psychological burden, and I am personally devoted to bringing it to an end for Sam and all type 1 diabetics."

After George W. Bush was inaugurated, Melton was summoned to
the White House, where he spoke to the president for about an hour.
At that time, the NIH's Human Embryo Research Panel, which included scientists and ethicists, had concluded that stem cell research
involving human embryos should be made available for research
supported by federal funds. But religious conservatives and antiabor-
tion activists were urging the president for a ban. Melton encouraged
Bush to view the matter the way President Kennedy had contem-
plated going to the moon — as something bold and visionary that
would define his presidency. He told Bush that he should focus on
"regeneration and repair" against these chronic, degenerative diseases. "That could be your moon shot," Melton said. "You could
change the whole health care system."

Bush ignored his advice, instead striking a muddled compromise.
On August 9, 2001, he announced that federal funds could be used
for research on embryonic stem cell lines — or those derived from
a single cell — created before that date, but not on lines created
afterward. The policy made little sense. If destroying a fertilized egg
is wrong — if it's the equivalent of taking a human life — it makes
no difference if the egg was fertilized before August 9, 2001, or afterward. Bush's promise turned out to be far less than advertised, anyway. Government scientists initially said that seventy-eight
batches of stem cells had been approved for study, but only about
twenty could actually be used. Among other things, they lacked genetic diversity, had chromosomal abnormalities, or had been exposed to "mouse feeder cells," which are derived from mouse embryos and are used to keep stem cells in their undifferentiated state
but may also be contaminated. If scientists were going to continue

the research, they would have to find their own embryonic stem cells.

In 2000, Melton went to a barbecue and met Douglas Powers, the scientific director at Boston IVF, a fertility clinic, and told him about his goal of turning embryonic stem cells into beta cells. He knew the clinic had unused frozen embryos, and he asked Powers if he'd be willing to share them. Until then, the clinic's surplus embryos were placed in orange biohazard bags and thrown out with the day's medical waste. But after consulting with ethicists, Boston IVF agreed to cooperate. The couples who provided the eggs still owned them, so they were sent four-page consent forms, asking that the eggs be released for medical research.

Once consent was received, a Boston IVF scientist would remove the plastic vials from the main storage containers, thaw the embryos, place them in a petri dish, put them in a portable incubator, and drive them to Melton's lab. In two years, 344 embryos were donated, allowing his team to tease out the stem cells.

This work, however, overlapped President Bush's announcement banning the use of federal funds on eggs fertilized after August 9, 2001. The restriction didn't stop Melton's research, which was funded by the JDRF and private donors, as well as the Howard Hughes Medical Institute. But special measures were necessary to ensure that no federal money was spent on new cell lines. A separate lab was created that contained no equipment bought with government dollars, which meant the lab needed separate centrifuges, computers, even light bulbs. Graduate students who were funded by the NIH were not allowed to work on those cells. Lab tools bought with private money were segregated into boxes with red stickers. "It was like making four dishes for dinner and making one of these with its own salt, pepper, and other ingredients," Melton says.

But it worked. By 2004, his lab had created seventeen new lines of embryonic stem cells, the most successful effort to date, doubling the number of lines available for research. Melton distributed cells to more than three hundred labs around the world, free of charge.

In 2001, Sam was an energetic ten-year-old who had recently gone on the pump. One day his sister, fourteen-year-old Emma, mentioned to her mother that she was thirsty all the time. Using Sam's meter, Gail tested Emma's blood sugar, and they stood around giggling as the numbers counted down. When she looked at the meter, the number was high. She tested again. Still high. She tried to keep her composure. There is no good time to get diabetes, but getting it as a teenager is probably the worst. Sam tried to be helpful. "If there is any way we can afford it," he told his mom, "you've got to get Emma a pump." (She did.)

Around this time, a young scientist from the Massachusetts Institute of Technology, Kevin Eggan, gave a talk at Melton's lab. Afterward, the two men sat in Melton's office to discuss a new experiment that was far more controversial than embryonic stem cell research.

While Eggan was interested in neurodegenerative diseases like Parkinson's, he and Melton shared a common frustration in investigating any genetic disease or one with a strong genetic component. The scientists didn't know exactly which genes were at fault, which thwarted their efforts to replicate the disease in genetically engineered mice, a mainstay in experimentation. The hurdle, however, could be overcome if the culprit genes could be identified, cultured, and studied — if a line of diseased cells could actually be grown. This would be markedly different from studying a fertilized egg; it would involve creating a human cell, which some equate to life itself. Melton had the raw material, the embryonic stem cells, to make the research possible; Eggan had developed the arcane skill of somatic cell nuclear transfer, more commonly known as cloning.

Cloning conjures images of mad scientists scheming to create Frankenstein's monster. In fact, virtually all scientists renounce "reproductive cloning" — that is, the creation of a new human being — as monstrous. What mainstream scientists like Melton advocate is "research cloning," which involves taking an unfertilized human egg and extracting its nucleus, which includes its genes and chromosomes. Then the nucleus of a donor cell, such as a skin cell, is transferred into the vacant egg, which is induced to divide until it becomes a mass of about two hundred cells, the so-called blastocyst stage. Researchers can then extract embryonic stem cells that are genetically identical to those of the original donor.

This approach, Melton says, could offer a window into the underlying nature of disease and could provide the basis for developing new drugs. For the first time, scientists could watch defective human cells, with their genetic permutations that cause the illness, develop from their embryonic state to maturity. If they could push the embryonic stem cells into becoming healthy beta cells, the cells could be transplanted into a diabetic donor and be accepted as "self," removing the need for immunosuppressant drugs. Melton says the experiment is vital even if it doesn't lead directly to a cure. "It speaks to the root cause of the disease," he says. "If we don't understand mechanisms, we're always going to be aiming for home runs. I hope someone hits a home run, but this is a very systematic approach to understand what's going on."

In 2004, Melton was named co-director of the new Harvard Stem Cell Institute, and later that year, he formally asked the university's review board for permission to produce cloned embryos for research. Cloning in the United States is legal but subject to institutional review. Melton's team would be the first to clone human cells in this country. "This is exactly the kind of work that we envisioned for the Harvard Stem Cell Institute," he told a reporter.

But his critics viewed it as a moral abomination — using science to create life for the purpose of destroying it. "I believe that Harvard is one of our premier research institutions, and that makes it doubly tragic that they are going down this avenue and sacrificing their moral credibility," said the Reverend Tadeusz Pacholczyk, director of education for the National Catholic Bioethics Center in Philadelphia. The governor of Massachusetts, Mitt Romney, said that cloning is unethical and should be illegal. When state lawmakers held hearings on the matter in February 2005, Maria C. Parker, the associate director for public policy of the Massachusetts Catholic Conference, a lobbying arm of the Catholic Church, testified that her father had died in a diabetic coma, but she insisted that he would have resisted treatments gleaned from the destruction of human embryos, comparing the research to experiments in Nazi Germany. "Science does not have to kill in order to cure," she said. The controversy caught the attention of the president, who said in his State of the Union address in 2005 that he will work with Congress to "ensure that human embryos are not created for experimentation or grown for body parts, and that human life is never bought and sold as a commodity." He received a standing ovation.

Melton works in a cramped Harvard laboratory whose location is kept secret for fear of protestors. A special key card electronically checks the identity of anyone who enters. Critics have attacked Melton personally, telling him that he's "doing the devil's work" and that "God will send you to hell." He lets them vent. "I don't ever say, 'Shame on you. My son and daughter have diabetes.' I never bring that up. I admit these are metaphysical questions, and I don't think they should believe what I believe."

But he and the institute defend their work. As its Web site points out, no fertilization occurs in nuclear transfers. There is no sperm, no implanting in the uterus, and no pregnancy. There are no body parts. "The unfertilized human egg, in my view, is like a tissue or an organ donation," he says. "It's not a person, it's not even a half a person." He bristles at Laura Bush's comment that "it really isn't fair to people who are watching a loved one suffer [to overplay the promise of stem cells.] We don't know that stem cell research will provide cures for anything." Says Melton: "You mean we're stuck with our lot in life and we shouldn't change it?"

He tries to expose his adversaries' inconsistencies. If life begins at conception, he says, what about miscarriages? The Catholic Church doesn't hold funerals for fertilized eggs that do not embed in the womb. "When I point this out," he says, "they find this to be a puzzle that they don't know how to answer. One answer is that this is God's will, and that's fine, but then that gets into this really complicated business of, is it then God's will to have a person like me wanting to work on human embryonic stem cells?"

The fundamental question — when does life begin? — will never be solved by empirical evidence or debate. It may be grist for philosophers or ethicists, but it's a diversion from alleviating real human suffering. If Melton must debate, he is more comfortable with mind-bending intellectual puzzles, such as what would happen if human stem cells were injected in a monkey's embryo. He would never conduct such an experiment, but he wonders if a human heart or brain would grow. "That is the kind of new biology that I find a million times more interesting than these specious arguments over whether life begins at fertilization," he says. As David Ewing Duncan writes, "Like many scientists, [Melton] likes to live in a space at the edge of what is possible."

Melton blames President Bush's restrictions for hampering his research, though he's been luckier than most. With funding from the Howard Hughes Medical Institute, the JDRF, and Harvard alumni, his work continues. Others in the field have been deterred, raising concerns that an area of enormous promise has been eviscerated or at least ceded to other countries. "If you really want science to succeed, it needs a bit of unfettered creativity," says Owen Witte, director of the UCLA Institute for Stem Cell Biology. "If you regulate it and restrict it and wrap it in chains, you are taking away the very essence of what science is supposed to do."

It is still unclear whether cloning cells will be possible. In 2006, Melton received approval from Harvard's institutional review boards, but new state or federal laws could block his work. He prefers to take the long view, noting that surgery was once considered abnormal, violating the body as the sacred vessel of the soul. Besides, he has more immediate concerns — like the actual disease.

What Melton has learned so far has made the cure seem even more remote. In one experiment, his lab found a flaw in a study that seemed to show that embryonic stem cells could easily be converted into beta cells. In another study, he concluded that beta cells are rarely if ever produced by adult pancreatic stem cells and that the pancreas may have no stem cells at all. This controversial finding contradicted the work of many other researchers, suggesting — if true — that their experiments are doomed to fail and raising the uncomfortable possibility that fertilized eggs will be the only renewable source of beta cells.

Stem cell research appeared to have made a significant gain in May 2005 when a South Korean scientist announced he had successfully cloned human embryonic stem cells, including one that was genetically matched to a diabetic. The work was hailed as the first step toward therapeutic cloning. But seven months later, the research was revealed as fraudulent: the investigator had faked at least nine of the eleven batches of cloned cells. Opponents of stem cell research cited the debacle as evidence that the field was controlled by renegade scientists. Melton says the discredited South Korean had only reported a technical accomplishment. The curative potential of embryonic stem cells, he asserts, remains unchallenged.

But Melton himself has certainly been challenged. He likes to say that a diabetic cure is a problem that can be solved; after fifteen years, however, his wife sometimes reminds him of his progress. "She'll say, 'You've been working all this time, and you haven't cured this disease,'" he says. "That's a lot on my mind." But of course he won't stop — his love for his children won't let him. He acknowledges, however, that his work has limits. While embryonic stem cells may cure other diseases, they will not alone cure type 1 diabetes. They may temporarily restore beta cell function, but they will not eliminate the immune system's attack on those cells. "If I'm wildly successful," he concedes, "I've only solved half the problem. So you might say, 'Doug, what sort of idiocy is it to work on half the problem?' But it's a recognition of how much one person can do."

CHAPTER 12

The Trials of a
Maverick Scientist

IN THE EARLY 1980s, Denise Faustman believed she would be part of a team to cure type 1 diabetes. As a postdoctoral student at Washington University, she helped develop the technique to isolate islet cells in mice and later participated in efforts to extract human islets. She extended her training by a year but finally left in 1985, when she joined the Massachusetts General Hospital for her internship and residency. Two years later, she became an independent investigator at the hospital, with a lab in Charlestown not far from the Bunker Hill Monument — a fitting symbol for the battles to come.

Faustman's charge was to create an islet transplant program to rival Paul Lacy's at Washington University. Despite her pedigree, she did not look the part. With a squeaky voice and a penchant for giggling during presentations, she seemed more like a cheerleader than a molecular biologist. It didn't help that Lacy treated her — in the words of one observer — "like a slightly goofy niece," making such remarks as, "Oh, she's so cute; she has a green thumb for doing science." Nonetheless, Faustman now had a chance to surpass her former colleagues and make Harvard Medical School, which is affiliated with Mass. General, the leader in this promising new field.

But a funny thing happened along the way. Faustman concluded that diabetes couldn't be cured through transplants, an opinion she reached by observing the continued rejection of the grafts. She also noticed that whole-organ pancreas transplants were far more successful in nondiabetics than in those with the disease. Why? Because while the transplants could temporarily restore beta cell function in diabetics, they didn't address the underlying problem: the immune system's attack on the islets. In other words, the transplants didn't cure the disease or even attempt to cure it. They simply remedied the symptom temporarily. Faustman decided that this entire line of research — developing a larger supply of islets (through better isolation techniques or through different animal sources) and inducing tolerance (through more sophisticated immunosuppressants) — was all a waste of time and money. "The problem is not graft rejection, it's disease recurrence," she said. While many of the investigators had noble intentions, she thought, the transplants had become a pipe dream sold to gullible financial supporters.

Faustman shut down her islet cell program and focused her work on the immune system, that inscrutable web of white blood cells, antibodies, toxins, and other chemicals that defend the body against attack by foreign invaders but can inexplicably turn on their own creator in autoimmune diseases. Faustman's move meant that she was abandoning a hot research field for something far riskier, which could imperil her own income. While her business card is inscribed with the impressive names of Harvard Medical School and Massachusetts General Hospital, neither institution gives her a dime in salary. She relies on grants, and grantors don't want to squander money on experiments with little chance of success.

Faustman still received money from traditional sources like the NIH and the JDRF, but had she relied exclusively on those, she could not have continued what she considered her most promising line of research. To do that, she found the academic equivalent of an angel investor: Lee Iacocca, the aging business icon whose wife, Mary, had died from diabetes — but not before he had promised her that someday, somehow, he would find the cure.

With Iacocca's funding, Faustman achieved the breakthrough she had hoped for — curing diabetic mice — and published the results in 2001 in the well-regarded *Journal of Clinical Investigation*. Curing mice per se wouldn't have been news (many researchers have done that), but Faustman's mice were the first to have had autoimmune diabetes (which more closely imitates the human variety), to have had "end-stage disease" (as opposed to newly diagnosed), and to have not required antirejection drugs. Her experiments also delivered unexpected results: the mice apparently regenerated their own beta cells, a finding so remarkable that the *JCI* editors refused to let her use the word "regenerate."

Faustman received approval from the Food and Drug Administration to conduct the first leg of a clinical trial that would potentially cure diabetes — an effort to translate the results from mice to men. Her work, as well as that of other scientists, contributed to putting regeneration in the mainstream. It no longer inspired perverse thoughts of salamanders regrowing severed legs. It is now called by its name, and it's become the next great hope for a diabetic cure.

On one level, Faustman's experience shows the extraordinary lengths that a contrarian investigator must go to for funding in a field — medical research — that favors incrementalism. Her own efforts included a personal makeover to improve her marketing appeal and charity events in the Missouri prairie; one of her supporters, meanwhile, tried to auction off four dead diabetic mice that she had cured. But the research has been eclipsed by the researcher herself. Faustman has attracted an army of dedicated fans, many of whom are parents of diabetic children, who believe that she alone speaks to their needs and frustrations. They have raised money for her through bikeathons, rallies, auctions, newsletters, and Web sites, turning her into the most famous diabetes researcher in America. Her work is also hailed by pro-life advocates, including some in the Senate, who claim her results preclude the need for embryonic stem cell research. As *Current Science* writes, "Denise Faustman is the Madonna of modern medicine. People fly across oceans to see her. Kids demand her autograph. Her email box bulges with messages."

At the same time, she is also the most controversial investigator in the field, a maverick whose flamboyant persona has never fit comfortably in the staid, male-dominated science world and who, by her own admission, is more of an "inventor" than a scientist. She became a doctor, she writes, "because I hated my pediatrician"; asked about her "biggest obstacle," she cites two: "guidance counselors at all levels of schooling [and] jealous peers who do not like changes in scientific dogma." Her critics include the most prominent names in diabetes — the JDRF, the head of the Joslin Diabetes Center, and the country's leading immunologists. Her harshest adversaries say her work cannot be reproduced and accuse her of outright fraud, prone to simplistic or outrageous claims that have given her a "cult following" among credulous patients. The animus runs so deep that the JDRF, which in theory should wish any diabetes researcher well, tried to undermine her work by distributing a scathing letter criticizing her.

Despite the attacks, Faustman has never been professionally reprimanded or disciplined, and her academic supervisor, Joseph Avruch, chief of the diabetes unit at Mass. General and a professor at Harvard Medical School, staunchly defends her, saying her critics are motivated by a combination of jealousy, sexism, and concern that her success could threaten their own funding. While Faustman herself is stung by the attacks, she says she has support from the right people. "The question is, who likes this work?" she asks. "The answer is, the people who have the disease. They say keep going, because if you don't, a cure's never going to happen, and we'll waste another twenty years."

Walk into any medical library to behold the scientific journals dedicated to this illness: *Diabetes. Diabetes Care. Pediatric Diabetes. Diabetes Technology & Therapeutics. Diabetolgia. Clinical Diabetes. Diabetes Spectrum.* That's only the beginning. Hundreds of medical journals include articles either on diabetes or on a related matter, such as obesity, nutrition, immunology, or any organ affected by hyperglycemia. Or walk through the exhibition hall of the ADA's annual Scientific Sessions to see rows of posters like stalks in a cornfield — each poster describing a different study. In Orlando in 2004, at least 1,400 posters swept across the hall, covering everything from macrovascular complications to integrated physiology to "gene chips and microarrays." There were twenty-six posters alone on diabetic dyslipidemia. Each year, thousands of researchers receive hundreds of millions of dollars to investigate diabetes; each year, thousands of articles are published; each year, biotech companies are created, breakthroughs are heralded, careers are advanced.

Patients are understandably frustrated. It sometimes feels as if they're merely fodder to be studied, source material for a vast medical research industry. They understand the complexity of the disease and acknowledge the exceptional therapeutic gains over the past three decades. But still they ask, With all this cutting-edge research, why haven't they been cured?

One answer, suggested by the microbiologist Lewis Thomas in his classic book *The Lives of a Cell* (1974), is that society is unwilling to pay for costly basic research, experiments that expand our knowledge as a necessary prelude to eradicating disease. Thomas believed that society instead contents itself with "halfway technologies," whose purpose is not to cure but to postpone death. These treatments are both "highly sophisticated and profoundly primitive," costing far more in the long run than developing and administering an actual cure. Organ transplants and many cancer treatments are good examples. So too is insulin. Patients still suffer with halfway technologies, Thomas wrote, but they are necessary "until there is genuine understanding of the mechanisms involved in disease . . . and the only imaginable source of this information is research."

But basic research is inherently risky, which is problematic for scientists who depend on grants for their salary and their lab. If their experiment fails, they will have difficulty publishing a paper, which will undermine their ability to renew the grant or receive another one. Their economic survival is far better served by conducting conservative experiments, whose results are easily predicted but whose scientific value is minimal.

"The academic investigator is in the business of progress," says David Scharp, a surgeon by training who spent twenty-five years in academic diabetes research, including many with Paul Lacy, and ten more years in industrial research. "If you make progress, then you maintain yourself and you get your publications and you get your grants. These are not necessarily end points, but you keep making progress. So as a matter of survival, you get dragged down to the status quo . . . If you're trying to eradicate a disease, that's not adequate."

These pressures, Scharp says, "have clearly slowed down progress in diabetes research because people aren't willing or able to take risks." Those frustrations contributed to his decision to leave academic research, and he is currently the chief scientific officer for

Novocell in Irvine, California, which is trying to develop encapsulation technology for transplanted islet cells. But Scharp says that commercial research, particularly for start-up companies like Novocell, has similar problems. Investors want to see immediate returns, yet biological research requires time — often more time than an investor will give. "If you have only nine months of funding, how do you mount long-term experiments?" Scharp asks. "It's hard, and that's why so many companies fail."

Scharp says that when he began his work with Lacy in 1972, he thought a cure was "totally achievable . . . And now it's 2005. That's ridiculous. And I look back and I ask, 'Why haven't we done it? Why hasn't it happened?'" The answer is complex, but what's certain is that much more work needs to be done in basic research. "What fails those of us who want to work in 'outcomes' research,'" he says, "is the lack of knowledge. That's where we always fail."

Research money itself doesn't predict success, because true breakthroughs often come more from serendipity than design — witness the discovery of penicillin. Alexander Fleming happened to notice that some mold had contaminated a plate culture of bacteria, and it had begun to dissolve. Fleming then grew the mold and found that it killed a number of disease-causing bacteria. This concept, a microorganism secreting something that would kill bacteria, became the template for all subsequent antibiotics. While luck played a role in the discovery, Fleming deserves credit for his observation and inquiry — testament to Louis Pasteur's comment that "chance favors the prepared mind."

The best minds should get the research grants, according to Michael Brownlee, the founder of the Diabetes Research Center at the Albert Einstein College of Medicine. "The people who make the breakthroughs have a track record, and they make them early," he says. But grantors rely on a business model approach that forces scientists to spell out what they will likely find over the duration of the grant. "If I already know what I'm going to get in five years, how important is it going to be?" asks Brownlee, who has type 1 diabetes himself.

Government funding for diabetes research has long faced specific hurdles. Infectious diseases in general receive priority over chronic conditions, and the perception that insulin is a kind of cure, or at least a potent remedy, has made other diseases seem far more threatening. As a result, the NIH, with an annual budget of $28 billion, has failed to keep pace with the diabetes epidemic. Since 1980, its budget for diabetes has increased by 240 percent, to $1.1 billion, but its total expenditures have grown by 261 percent. Thus the percentage given for diabetes has declined slightly even as the number of diabetics has doubled. As the *New York Times* determined, the NIH in 2004 spent $68 for each diabetic, compared to $16,936 for each patient with the West Nile virus. And the news gets worse. In 2007, the NIH will actually reduce its spending on diabetes research by $1.2 million.

Ronald Kahn, at Joslin, has tried to calculate the government's neglect. With at least 20 million diabetic Americans, public spending is at most $50 per person. But the average cost of care per diabetic, according to Kahn, is between $10,000 and $11,000 a year — so the $50 expenditure is a pittance. "Our investment in the future, the future of all these people, amounts to less than half of one percent of what we're spending on the disease," he says, noting that even the tire industry spends at least 3 percent of its sales on research. "It's simply not enough. I can't say that if we invested ten times as much we'd move ten times faster, but we would move faster."

The NIH is by far the largest noncommercial funding agency of biomedical research in America, but how it makes specific grants is another problem. In a commentary expressing concern about innovation in research, the medical journal *JAMA* says the NIH needs to support high-risk, high-reward studies. According to *JAMA,* NIH's study sections, which review and recommend grants, often turn down new researchers because their applications do not contain enough preliminary data to confirm the feasibility of the proposed work. "If a young researcher has had superb success in his or her training . . . and if he or she proposes exciting but unproven experiments, that researcher should be given the resources to move ahead," *JAMA* says.

The NIH concedes the funding problem. The director of its Center for Scientific Review, Toni Scarpa, writes in *Science* in January 2006 that the NIH is funding a smaller percentage of grant applications, intensifying unfavorable trends. "Competitive pressures have pushed researchers to submit more conservative applications," she writes, "and we must find ways to encourage greater risk-taking and innovation and to ensure that our study sections are more receptive to innovative applications."

Denise Faustman vividly recalls treating patients in the diabetes clinic at Mass. General. One woman in particular — she took care of foster kids — stands out. "She was sitting down, talking to me," Faustman says, "and I picked up the bottom of her foot, and she had this crater with pus in it. She was already partially blind, and she had had a stroke. It was on that day I said, 'This is ridiculous. What am I going to do? What impact can I have on her life? I'm just going to follow the deterioration.'"

She already knew what she wanted to do. "She had a clear-cut plan," says her supervisor, Joe Avruch. "She didn't want to engage in patient care. She wanted to cure type 1." Given her credentials, Avruch considered her a find, and in 1991 his faith was rewarded in what he called a "stunning" achievement. Though only a thirty-three-year-old junior faculty member, Faustman had published two papers in *Science,* one of the country's top medical journals. The first presented a novel theory that in children at high risk for diabetes, certain white blood cells had on their surface reduced levels of correctly assembled MHC Class 1 molecules. These molecules are supposed to display tiny bits of "self" proteins to the immune system's killer T cells, essentially teaching them to ignore the body's own tissues. According to Faustman, instead of containing the protein fragment — also known as peptides — many of the molecules were empty. This work was highly controversial because it defied decades of genetic studies suggesting that type 1 diabetes was associated with a single abnormality in the portion of the genome responsible for encoding another MHC molecule, Class II, which displays proteins from invading microorganisms. Though Faustman says her results have been confirmed by other researchers, some of her peers continue to dispute her findings.

Faustman's second article only intensified her reputation as an iconoclast. Seeking to promote the tolerance of islet cell transplants, she ignored the conventional approach — treating the recipient with antirejection drugs — and decided to treat the tissue instead. She coated human islet cells with chemicals to cover a protein that the immune system uses to identify invaders, then placed those cells, not in another human, but in mice. This transplant struck some scientists as bizarre; nonetheless, the transplanted cells did not provoke any immune response, raising the possibility that the masking technology could be used to transplant islets into humans without immunosuppressants. This experiment was covered in the *New York Times* and other mainstream publications. While Faustman cautioned that the research was at least five years away from testing in humans, she described the complicated science in an exciting though somewhat mystifying fashion. "We're making invisible tissue," she said.

The masking technology, however, was never translated to humans — an indication, her critics contend, of a researcher whose initial findings make a big media splash but never live up to the hype. Faustman soon concluded that islet transplants would fail because the diabetic's underlying autoimmunity would kill the graft. She wanted to investigate what actually caused the disease in order to cure it. Fortunately, her efforts caught the attention of a medical adviser to the Iacocca Foundation, which would become her most significant source of funding.

Lee Iacocca met Mary McCleary in 1948, when both worked at the Ford Motor Company — he had been hired two years earlier as an engineering trainee, she was a twenty-two-year-old receptionist. Mary was diagnosed that same year, and her struggles with volatile blood sugars, particularly hypoglycemia, began immediately. "From the day she had it, it was bad," Iacocca says. But her health didn't deter his courtship. The couple married in 1956, just as Iacocca's own career took off. He recommended offering the 1956 Ford for a modest down payment, followed by installments of $56 a month for three years. The "56 for 56" was his first marketing coup. He later introduced the Mustang. After his dismissal from Ford, he was hired by the moribund Chrysler Corporation, which he famously rescued in part through his commercials as the straight-talking executive who vouched for the quality of the cars. The 1980s was an era of celebrity CEOs, and no one played that role better than Iacocca — the son of poor Italian immigrants who could intuit what American consumers wanted in their vehicles, a patriot who raised $540 million to restore the Statue of Liberty and Ellis Island, a man of the people whose ghostwritten autobiography would sell 7 million copies.

If Iacocca's life reflected the American dream, then diabetes was its recurring nightmare. His wife wasn't his only exposure to the disease. He saw that diabetic employees often couldn't work an entire day and that they were ostracized in the plants — perhaps because the needles were associated with illegal drugs. Diabetics "were a cut above lepers," he says. Iacocca also knew the brutal realities. Mary had three miscarriages, though she was able to deliver two daughters by caesarean section. Her hypoglycemic experiences in the dead of night were harrowing — her stiff body breaking out in cold sweats until the paramedics arrived, the red light from their ambulance flashing through the window, the girls cowering in fear. When Iacocca traveled, he called Mary two or three times a day; he could tell by her tone if her blood sugar was low. He hired someone to spend the night in case she slipped into unconsciousness.

Nothing could bring Mary under control. She would fly to Boston for care at the Joslin Clinic, and she briefly wore an experimental insulin pump. She nixed it, however, because she feared it would go haywire and kill her in her sleep. She had her first heart attack in 1978, three months after Ford fired her husband. It was the beginning of a pattern: over the next four years, she would have another coronary and a stroke, each event following a tumultuous period in her husband's career. This convinced Iacocca that stress contributed to her failing health, and he would hold himself partly responsible for her demise. But Mary's spirit remained strong, and she refused to complain or feel sorry for herself. "You think I have it bad?" she would say. "You should have seen the people who were with me in the hospital."

For all his money, for all his power, Iacocca could do little more than watch his wife waste away. During one hospital stay, a doctor suggested amputation. Iacocca said no, telling him, "She can't wake up without a foot." Mary was adamant about one thing: she wanted a cure. Not for herself, of course; she knew it was too late for her. But when she went to Joslin, she saw the children sitting in the lobby and running down the halls; she wanted it for them. Her husband promised that he would find the cure.

In 1983, Iacocca appeared on TV to pitch K-cars with his signature line: "If you can find a better car, buy it." That same year, Chrysler paid back its federal loans, marking a high point in Iacocca's career. And in 1983 Mary Iacocca died; she was fifty-seven. "Her tired heart just gave out," Lee says. "She was still very beautiful." At her insistence, her death certificate cited "diabetes" as the cause.

The following year, the Iacocca Foundation was established to fund basic research, and over the next twenty years, it distributed more than $23 million in grants. Iacocca was confident that his far-flung team of "scientists in white coats" would make the necessary breakthroughs. "At one time I owned 5,000 mice," he says. "I used to have 5,000 cars in inventory, but the cars I could sell. What do I do with the mice? I give them diabetes and kill them."

In 1999, the foundation's board members listened to a presentation from one of its promising scientists, Denise Faustman. By then, it had given her $5 million. She described a recent paper in which she had identified in nonobese diabetic (NOD) mice* a defective "protein cell pathway" in certain white blood cells, or T cells (so called because they mature in the thymus gland). This regulatory pathway controls several immune system activities, but Faustman said she was the first to link it with autoimmune diabetes. The discovery allowed her to use certain drugs to kill the pathogenic T cells in tissue culture, which could potentially be done in diabetic mice, without immunosuppressants. That would require another experiment.

At the end of Faustman's talk, there were no questions, just silence. Faustman feared she had used too many medical terms and assumed her presentation was a flop. Then Iacocca looked at her. "Denise," he asked, "why aren't you doing the experiment?"

She stumbled for an answer, then started to list the reasons. First, she said, the experiment would cost half a million dollars. Second, it would probably fail, as mice with end-stage diabetes have never been cured without immunosuppressants. Third, she would have a hard time hiring postdoctoral students to work on an experiment that

* These mice have autoimmune diabetes, so their disease is considered to most closely resemble type 1 in humans.

probably would not produce a paper. Finally, a failed experiment would set back her own career at Harvard.

Iacocca didn't care. "I'll give you the money," he said.

Faustman wasn't convinced. If the experiment failed, she asked, would the foundation still support her?

"I don't care what the result is," Iacocca said. "Just do it."

Faustman later said, "It was a dead-end experiment for my career, and it was a dead-end experiment for the people who work with me. You don't want a three-year gap on your résumé that says you didn't do anything, and you don't want to write a paper that says you spent half a million and got zero data."

But with Iacocca's support, she did the study. As a safeguard, she had her postdocs conduct a second, predictable experiment that would give them an easy paper. In the principal study using NOD mice, Faustman's premise was that the genetic defect in the autoreactive T cells directed their assault against the beta cells but also made them susceptible to a cytokine, or signaling protein, known as TNF-alpha. (Its full name — tumor necrosis factor — is derived from its ability to kill cancer cells.) Ironically, TNF-alpha *antagonists* (which destroy the cytokine) have been used to treat diseases, such as inflammation, but Faustman wanted to use TNF-alpha *agonists* (which promote it) as part of a diabetic cure, an approach, she says, that had never been tried.

Her experiment required two steps. First, she injected the mice with a compound known as Complete Freund's Adjuvant (CFA), triggering the TNF-alpha. Second, she tried to "retrain" the immune system by injecting the animals with healthy cells taken from the spleens of other mice. These normal cells were to present the self-peptides that instruct the immune system to attack only harmful invaders, not healthy tissue. Faustman hoped this would deter the immune system from making the defective T cells.

Even if successful, the treatments would stop only the auto-immune attack. They wouldn't lower blood sugars — the mice still needed islet transplants. But a transplant without antirejection drugs would still represent a huge victory.

The work itself, Faustman says, was "a horrific experience." The animals were kept alive through insulin injections or islet transplants, their blood sugars measured twice daily. After a year and a half, positive data began to emerge: about 80 percent of the animals achieved normal glycemia, indicating that the transplanted islets could survive without immunosuppressants. "There was jubilation in the lab," Faustman says.

She could have ended the study there, killing the mice — there were fifty — closely analyzing the pancreata, and writing a paper about her lab's success in reversing autoimmunity in NOD mice. But she wanted to do one more experiment — remove the islet capsules, lodged beneath a kidney, to record the elevated blood sugar that would result. Her two postdocs and one assistant said the results would be self-evident and that the pancreata could be damaged during the surgery. The work would also cost more money, but Iacocca's support allowed Faustman to continue.

Her staff began the final experiment. The morning after the kidneys with the islet capsules had been removed from two of the mice, her two students walked into her office, their shoulders slumped. They had tested one of the mice.

Faustman was eager to hear the news. "What's the blood sugar?"

"It's normal. One-seventeen."

Faustman was angry. She thought her students had operated on the wrong mouse or removed the wrong kidneys, or maybe the glucose meter was broken, or perhaps the animal wasn't eating yet. She ran to the cage and saw it running around, well fed and healthy. She went to Avruch and told him she couldn't figure out what was happening. A pathology report on the kidney, showing the islets, confirmed that the correct organ had been removed. A week passed, and the mouse, as well as a second one, maintained normal blood sugars. Where was the insulin coming from? They killed the animals and examined the pancreata, and where before there had been no islets, there were now — in Faustman's words, "big, fat islets. It wasn't a subtle finding." In all, seven out of nine NOD mice that had received CFA and splenocytes maintained normal blood sugars for more than forty days after the transplanted islets were removed, and the islet mass after at least eighty days of disease reversal was about 50 percent of its original value. It appeared the animals' own islets had regenerated.

Faustman wrote her paper, but medical journals were skeptical. "Getting it published was extremely hard," Avruch recalls. "No one could believe that she had permanently cured an NOD mouse in an advanced stage of diabetes." Faustman offers a fuller explanation of the skepticism. "You're born with one brain. You're born with one heart. You're born with one pancreas," she says. "Maybe your skin regenerates. Maybe your hair regenerates. But your organs don't regenerate. At least that was the thinking." The *Journal of Clinical Investigation* published the paper in July 2001 without using the word "regeneration." Instead, Faustman wrote: "suppression of hyperglycemia . . . promotes the functional restoration of endogenous B cells or their precursors."

Next, Faustman wanted to determine if the islets regenerated from another source or if they were "rescued" from the pancreas once the autoimmune attack halted. She transplanted spleen cells from healthy male mice to "reeducate" the immune cells of female diabetic mice, and she found that all of the new functioning islets had significant numbers of Y chromosomes, indicating they had come from male donors. In another experiment, donor spleen cells were marked with a fluorescent green protein; again donor cells were found through the new islets, raising hopes that the lowly spleen had a curative potential for diabetics. But a separate experiment indicated that the islets could also grow from the recipient's own remaining precursor cells, resuming insulin secretion once the autoimmune process had ended. Spontaneous regeneration now seemed possible. Faustman published her results in *Science* in November 2003, this time using "regeneration."

The following year, the JDRF held a regeneration workshop, identified regeneration as one of its six "therapeutic targets," and pledged $10 million to the cause. While other scientists were also making important contributions, Faustman received most of the publicity through the Iacocca Foundation, through trade publications like *Diabetes Health,* and through the *New York Times* and other mainstream outlets. In May 2005, *O, The Oprah Magazine* featured Faustman as making one of the "five biggest health breakthroughs by women scientists" in the previous five years. Faustman said the discoveries would have never happened without Iacocca's support because "the system is not set up to ask hard questions."

Alfred Hershey, a scientist who won a Nobel Prize in physiology, described heaven as coming to work every day, doing the same experiment, and always getting the same result. This notion, dubbed "Hershey heaven," reflects scientists' pursuit of truth, any truth, as a paramount virtue; but for others — including patients hoping for a cure — it highlights a fetish for predictability at the expense of progress.

Denise Faustman had her own idea of heaven. After her successes, she could have done more animal studies. As her critics pointed out, her experiments lacked two controls: mice not receiving CFA and mice not receiving islet transplants. Which treatment was critical? Most scientists, in fact, would have done more studies to ensure their accuracy and their reproducibility, and to gain a better understanding of the mechanisms behind the observations. But Faustman prefers pushing forward to the next experiment. Even her supporters, including her boss, say they would not copy her approach.

"I like to do controls over and over," Avruch says, "but she doesn't want to be bothered. Her style is to make inductive leaps that often don't seem to be well supported by the data, and then she moves on from where she lands. Did she jump for the right reason? Who knows? But what's happened subsequently is very impressive. Unfortunately, that's not the style that impresses card-carrying investigators." Her approach makes her "more of an inventor than a scientist, like these guys in their garage, tinkering with new ways to do things, but she does them with a tremendous degree of molecular and biological sophistication. And as long as she gets results, that's fine."

Faustman, who concurs with the "inventor" analogy, saw no reason to retrace her steps. She had already cured mice. Now she wanted to test her experiment in humans, conducting a clinical trial that would cure diabetics. That may not sound remarkable, but it truly is. Clinical trials of any kind are expensive, time-consuming, and risky, with relatively few qualified investigators; a trial that would cure any disease is rare.

Faustman received approval from the Food and Drug Administration to conduct the first leg of her trial, but she needed at least $11 million. Iacocca was still committed to the research, but he couldn't fund it himself. He thought he could raise the money by asking his richest friends. When that didn't work, Faustman herself hit the road.

For money and respect, medical researchers have to market themselves, discussing their work with other scientists at conferences, professional meetings, and social events. Their peers' perception of them is important because the peer review system for grants — at the NIH, for example, or in diabetes, the JDRF — means their colleagues will largely determine their economic fate. Professional marketing is important for another reason. Given the obscure nature of their work, the media will ask their peers to explain and assess their research, commentary that is particularly important in Faustman's specialized field. Few if any mainstream journalists, even those writing on science, could decipher a research paper on immunology or molecular biology. Even other scientists would be hard pressed to assess its significance; only others in that field have the knowledge.

Another type of marketing relates to fundraising, as researchers must convince would-be donors to contribute. If the audience is already conversant in the research, the scientist doesn't have to translate the work. But if it's a lay audience, scientists must walk a fine line. If they use the language of the lab, few will understand it; but if they "dumb down" their research, they risk misleading the audience by oversimplifying complex ideas.

Faustman never really cared about marketing herself. She didn't reach out to her peers, she said, because she believed her data were sufficient. Besides, she wasn't very good at it. As Avruch says, "She had this bubbly presentation, and she looked like a cheerleader. [In science] that's not insignificant." According to Dana Ball, executive director of the Iacocca Foundation, Faustman looked "too casual and unsophisticated . . . She laughed too much."

When it became clear that Faustman needed public support for her trials, Ball persuaded her to change her appearance. She got a haircut, bought new glasses, wore conservative suits, put on earrings. Faustman shrugged off the changes, saying that "long, flowing hair wasn't me" and that she tired of "trying to fit in with the Birkenstock crowd." What remained was the best part of her personality — a natural warmth and empathy that defied the stereotype of the arrogant, isolated scientist. Faustman also worked closely with a publicist in New York, Russell LaMontagne, who was initially hired by the foundation but was then retained by Faustman herself. LaMontagne arranged publicity events, helped science reporters craft favorable stories about his client, and honed Faustman's message. Her presentations became more tightly scripted, more serious, and while she'll never project gravitas, she won't be confused with a cheerleader either.

A large audience, in fact, was ready to embrace her. Part of her popularity came by default; those hoping for a cure believed they had nowhere else to turn. The JDRF, which has allocated more than $900 million for research, is currently funding clinical trials for prevention, complications, and reversing the disease in new patients — all worthy goals, to be sure — but no trials for patients with established diabetes. What about the children? Embryonic stem cells are a quixotic dream. An artificial pancreas is a distant, high-tech, halfway solution. And proponents of islet cell transplants have concluded — as Faustman did long ago — that new islets will not cure the disease any time soon. Camillo Ricordi, the scientific director of the Diabetes Research Institute at the University of Miami, who developed an automated procedure for isolating islet cells, said that the goal of current transplant trials is not "insulin independence" but to improve "the autoimmune response" and "to maintain islet mass and function." Indeed, no responsible parent would subject an otherwise healthy diabetic child to a lifetime of toxic immunosuppressants. Nonetheless, the NIH announced in 2004 that it would spend another $75 million on islet cell research, and the JDRF continued to hail the Edmonton protocol as a success — "It works," Peter Van Etten, president of the JDRF, assured *Forbes* in 2004. At the 2005 fundraising gala of the JDRF's New England chapter, the organization designed a brochure that equated the protocol to the discovery of insulin!

Speaking in Boston in March 2005, Van Etten said that funding for islet cell transplants reflected the JDRF's primary research goal.

"It's important to step back," he said, "and consider that the goal here ultimately is — as important as it is to throw away the insulin, throw away one's syringes and insulin pumps — the primary goal is to stop the onset of complications, because it's the complications that are the threatening aspect of this disease."

Complications are indeed harrowing, but for many parents, avoiding them is not enough. They want their children to wake up in the morning, get dressed, eat breakfast, and go to school — without injections or infusion sets or test strips or continuous glucose monitors or sugar tablets or low-carb bread or immunosuppressants or stainless steel bracelets that mark their children as different, diseased, and disabled. They want the cure.

They know that picking which scientist might find it is impossible, but if there's only one horse in the race, that's the one they'll bet on.

In the early 1970s, when Lee Ducat's son was diagnosed, she had to go the library, page through medical journals, write letters, and make telephone calls — only then could she create a movement to find a cure. But technology has changed the speed and power of grassroots campaigns. When Christina Smith's daughter, Gracen, was diagnosed in 2003, Smith sat down at her computer, Googled "diabetes," and began reading. She found a handful of scientists and e-mailed them, asking for more information. She communicated with researchers in the United Kingdom, Israel, and Singapore. Denise Faustman eventually e-mailed her back, sending her a newsletter as well. Tapping into the Iacocca Foundation's Web site, Smith could communicate with other parents around the country who had studied the research and were willing to raise money for Faustman.

Smith, who lives in the small town of Maryville, Missouri, tucked away in the northwest corner of the state, wanted to meet Faustman, so she flew to Palmdale, California, where the scientist was appearing at a fundraiser. Smith met Lee Iacocca as well. In October 2004, the Iacocca Foundation asked Smith if she would appear with him on the *Today* show to commemorate the foundation's twentieth anniversary. Smith flew to New York, carrying a poster with photographs of her children, the foundation's Web site address, and the phrase

"Believe in a Cure." While she wasn't interviewed, the poster made it onto the show.

Smith decided she would hold a fundraiser for Faustman and the Iacocca Foundation, and the two principals agreed to meet in Missouri.

Others were raising money for Faustman as well. Two women in Connecticut, Susan Root and Jackie Fusco, both of whom have diabetic children, sponsored an annual bike ride, "100 miles for Human Trials," from Mystic, Connecticut, to the Mass. General Hospital in downtown Boston. The event raised $145,000 in 2004 and $190,000 in 2005. A group in Hendricks County, Indiana, spent four years raising money — through charity motorcycle rides — for a rare neurological disorder called Batten disease, which affected two boys in their community. The fundraising helped the boys receive life-saving gene therapy treatment, so the group, hoping to conquer another disease, chose juvenile diabetes and decided that Faustman was their only real hope for a cure. With the motto "You don't need to ride. You just need to care," the "Dallas and Reid's [Motorcycle] Ride" raised $54,000 in September 2005. That same month, the nonprofit DOCS (Diabetes on the Cure in Seattle) Foundation raised $12,000 with a motorcycle ride and concert, which included performances on the cliffs of the Columbia River by Styx, Foreigner, and REO Speedwagon.

Now it was Christina Smith's turn. Armed with pictures of her daughter, she went to diabetic support meetings, hospital meetings, Rotary Club meetings, retired teachers' meetings, Kiwanis meetings — any meeting at all — telling them about her daughter's disease, Faustman's efforts, and the coming fundraiser. She sought out corporate sponsors for food, libations, decorations, and items to be auctioned. Nothing came easily. She secured one restaurant but had to cancel it because the event's size would have violated its fire code. She thought she had reserved another locale until its owner presented her with an estimated invoice for the facility, decor, plates, napkins, and security. She had convinced Tom Watson, a golfing legend and a Kansas City native, to attend a cocktail party the night before the event, and guests could pay a fee to meet both Watson and Iacocca. But it was canceled because Iacocca had to be in Dallas for a wedding.

Called the "Bling Bling Ball," the fundraiser had a diamond theme: diamond sponsors, diamonds printed on publicity materials and tickets, and a Diamonds for Diabetes raffle. Guests bought $20 glasses of champagne with a number attached. At the end of the night, the winning number would get the diamond. Unfortunately, the jeweler who was expected to donate the stone declined at the last minute. Smith thought about donating a diamond she owned, but instead used a gemstone called ametrine, donated by a diabetic jeweler.

Iacocca donated a case of wine from his vineyard in Italy and an autographed copy of a 1985 *Time* magazine cover of himself, headlined: AMERICA LOVES LISTENING TO LEE. But while an entire nation may have loved listening to Lee twenty years ago, his audience today is smaller. Smith tried to sell his table for $50,000 but got no takers. Then she offered it for $25,000. Still no bidders. So she had wealthy guests sit with him, hoping that they might contribute more than the $100 admission fee. Smith also hoped that the famed investor Warren Buffett, whose estimated worth is $44 billion, might attend. A friend of Iacocca's, he lives ninety miles away in Nebraska. Guests surely would have paid to sit with the Oracle of Omaha — perhaps $150,000, Smith estimated. But the week of the event, Buffett was being questioned by government authorities about the insurance industry, including practices involving one of his subsidiaries. Buffett didn't make the Bling Bling. A friend of Smith's e-mailed Bill and Melinda Gates, whose foundation promoting global health has an endowment of almost $29 billion, but got no response.

A Kansas City news station was to cover the fundraiser, but Pope John Paul II died that day, so the Bling Bling got bumped.

Iacocca himself almost didn't make it. He prefers nonstop flights in first-class seating, but none is available from Dallas to St. Joseph, Missouri. Smith called her congressman, Sam Graves, whom she had met in Washington to discuss Faustman's research. Graves directed her to the company in Kansas City with whom he had chartered flights, and Smith secured a plane, paid for by Iacocca.

The event itself takes place on the green rolling hills of the St. Joseph Country Club, whose eighteen-hole golf course with zoysia fairways and blue-grass roughs is the pride of northwest Missouri. It's a gorgeous spring evening, and a black-tie crowd sips cocktails, listens to a pianist, and reviews the auction items, including two NASCAR tickets at the Kansas Speedway, a model of a 1965 Ford Mustang, and, of course, Iacocca's Italian wines. Faustman, who is divorced, flies from Boston to Kansas City, leaving her two children with their nanny, then drives forty-five minutes with Ball and LaMontagne to the event. When she arrives, few people notice her. "I'd rather be nerding out in the lab," she confides. Most are waiting for Iacocca, whose plane lands shortly before dinner. No longer the swashbuckling CEO, he walks stiffly and gingerly, smiling for photographs, sipping a drink, and shaking hands with well-wishers.

Helping to organize the event is Debra Hull, whose fitted black dress could not accommodate her insulin pump — it's strapped to her thigh — and whose lovely rhinestone necklace leads to a black teardrop in the middle. "I needed a way to express the pain of this disease," she says, "without putting a damper on the evening." Diagnosed when she was nine, she developed retinopathy by nineteen. She became a nurse and was also a sales representative for an insulin manufacturer. Married, with three adopted children (pregnancy was too risky), Hull is now thirty-seven, a farm wife who works at a hospital part-time to receive health benefits. She's in good physical shape, but she is "haunted" by her circulation problems. "You can probably hear my arteries hardening," she says, adding that she feels "guilty because I let this happen to me."

She is also angry: at the doctors she used to work with, who are too lazy to teach their patients or too fearful about getting sued if tight control leads to hypoglycemia; at the government and private insurers, for failing to reimburse patients for their care; and at society in general, for making diabetics feel guilty. "Go to any funeral of a diabetic," she says, "and you'll hear at least one person say, 'He never really took good care of himself. I saw him eat an ice cream cone back in 1984.'"

She is also angry at the financial cost, which she estimates has been $180,400 in her own case. That includes $46,100 for test strips and, before that, urine-testing supplies, $31,925 for insulin pumps and pump supplies, $6,000 for a car accident from hypoglycemia, and $800 for alcohol swabs. Because she lives in the country — more than fifty miles from the nearest endocrinologist — she adds an-

THE TRIALS OF A MAVERICK SCIENTIST

other $8,500 in gasoline for all of her doctors' appointments, lab tests, pharmacy needs, and hospital visits.

Like many diabetics, she carries on with the help of her family and her faith, citing one passage in the Bible from Romans 5:3–5: "Rejoice in our sufferings, because we know that suffering produces perseverance; perseverance, character; and character, hope. And hope does not disappoint us, because God has poured out his love into our hearts by the Holy Spirit."

But for all her years as a patient, nurse, and educator, Hull has never supported any one scientist. Survival has been a full-time occupation. "We're too broke and too tired to fight for a cure," she says. And when she did invest her hopes in a cure — islet cell transplants at one point, the encapsulation of cells at another — she was always disappointed. But her attitude changed when Christina Smith passed out a flyer at her church and spoke to her about Faustman's work and about Gracen. It struck a chord. "At first, you're too busy to think about a cure," she says, "and then you find empty promises, and then you decide, well, it's too late for me anyway. Now you just hope that no one has to go through this again."

After dinner, Hull steps to the microphone to introduce Iacocca. She had written a speech in which she had described diabetes as "a terrible beast chasing you," but Iacocca's representatives rejected it, saying it was too negative. So Hull says she wishes for the day that "diabetes will be a thing of the past."

Iacocca, enjoying the spotlight, quickly warms up his audience. "I hear this is a big turnout for St. Joseph," he says. "I thought I'd be speaking in Kansas City, but I spent most of my night speaking to people from Maryville, wherever that is."

He lays out the challenge of raising money. "I did it for the Statue of Liberty and Ellis Island, and we raised $540 million. But patriotism and finding your family roots is an easy sell. Diabetes is a tough sell. Most people think you take a couple of needles a day and everything is okay. But your eyes go and —" He stops and looks around the room. "You have to watch what you say around young children. You don't want to get too morbid about it."

Iacocca says he's given many speeches on diabetes, but tonight he will do something different. In 1979, he received an award from the

ADA, and that night he gave a speech in New York about his hopes for curing his wife. The speech is still relevant today, he says, "because nothing is ever new in this business." He pulls the speech out of his pocket and reads a shortened version, which includes his gratitude to the health care professionals and scientists he had met over the years. "And then I've come to know many of the most important people of all, the people who wage their own personal battles against diabetes every day, every hour, and every minute of their lives. These are the people like my wife who really have something to teach us all. They make abstract concepts like courage and discipline and determination come alive every day, and they show us what the human spirit can endure and accomplish . . . We're going to find the key or keys to get rid of this terrible disease, and we hope and pray we find it soon."

Iacocca looks up to the crowd. "So, that was twenty-six years ago. Where are we now? Still no cure. Twenty years of working and waiting, still no cure. So am I discouraged? Well, yeah, a little. Am I frustrated? Who wouldn't be? But have I given up? Absolutely not, and I'm going to keep fighting until we win this thing. Now I'm going to sit down and listen to Denise Faustman. She's a good girl."

Iacocca and Faustman stand for pictures, and after an ovation, she begins her remarks. Her appeal is easy to understand. Reading from note cards, she rarely uses the language of molecular biology but relies on straightforward ideas. After she realized that transplants were futile, she says, she returned to the lab to investigate "the fundamental question of why do people get type 1 diabetes." There were, however, "three dogmas from the research community. The cure comes from prevention. Focus on complications. And the cure comes from genetic efforts." But these truths have led to disillusionment among diabetics and their loved ones. If their goal is to be cured, she says, the first two "dogmas" won't help them, and "type 1 defies the rules of genetics," as studies have shown that if one identical twin gets the disease, the other one develops it only 40 percent of the time. Instead of studying genes, she says, she investigated the proteins in the blood cells from diabetic humans and mice. "We found major protein differences between the good white blood cells and the bad white blood cells," she explains. "And that was good because the bad white blood cells kill the insulin-secreting cells in the pancreas." She asks the audience how "protein discovery work" could help humans, and she answers her question with a simple analogy. "Take antibiotics. They kill the bad bacteria, but they don't kill you." That's possible, she says, because the antibiotics can target the proteins in the bacteria. This is what her lab tried, designing "compounds to kill off the bad cells."

Faustman says the experiment had a low chance of success — "Who would try to reverse diabetes in mice when it's never been done before?" — and was only possible through Iacocca's support. The experiments took three years, but after treating the diabetic mice with two compounds, most emerged with normal blood sugars. "We thought we had reversed end-stage autoimmunity," she says. She then explains how the lab made its surprising discovery of islets in the diabetic mice. "There was spontaneous regeneration," she asserts.

It has been, until now, a stripped-down tale of a scientist's journey through an immunological minefield, but its conclusion raises new and unexpected possibilities. "A natural healing process occurs" in diabetic mice, Faustman says, "as cells from the spleen come across and regenerate into islet cells. Maybe the diabetic has this amazing regenerative potential and capability. All the animals needed was the disease removed, and the process took over by itself. We don't think that you have to transplant islet cells at all. We think that the pancreas has the inherent ability to regenerate, and that the problem with diabetes is not who's got the bigger vat of stem cells, the bigger vat of islets, or who's got the bigger herd of pigs for islet donations. We think you can do it yourself if we can get rid of the disease."

She closes by citing her patron. "As Mr. Iacocca likes to say, if you're a mouse, I got you covered, but that's not the big story, is it? The big story is whether we can cover people too."

After the applause, the Reverend Sidney Breeze walks to the microphone and says that his wife died from diabetes, "but with what we know now, she would be alive. So I compliment Mr. Iacocca and Dr. Faustman for their efforts, and our efforts tonight. Let us pray."

Faustman never promises a cure or suggests a timetable, but what she proposes is something even better — a remarkable act of self-empowerment. Diabetics live their entire lives at the sufferance of some pharmacological or medical intervention, existing on permanent life support, wards of the medical-industrial complex. Faustman says, If I'm correct — if I can "kill off the bad white blood cells" — you will heal yourself. It's an intoxicating message delivered with humility and charm. Little wonder that Iacocca says, "I wish I had known her when I was selling cars and trucks. She could sell anything."

After her talk, Faustman's new supporters shake her hand, hug her, and thank her for her work. "I'll be praying for you," one woman says. Another has tears in her eyes. A long line waits to take commemorative photographs with her and Iacocca. "It better be a good hair day," Faustman quips. She returns to the banquet room, where the three-piece band, Exit 47, is playing. Smith's seven-year-old son, Paden, asks her to dance, and the two are soon twirling on the floor to "Kansas City, Here We Come." Little Gracen, beautiful in a black dress, dances too.

The night wears on, and Iacocca, exhausted, prepares to leave. Since Mary died, he has been twice married and twice divorced. His is a rich and comfortable life, but he knows his time to make good on his promise is running short. "I just reached the magic age of eighty," he says. "I always said, 'Give me five years.' But now I'm running out of five-year cycles."

The event raises $60,000, far less than the $150,000 that Smith had hoped for — and small change compared to the million-dollar galas of the JDRF, which auctions off $6,000 dinners for ten at fancy restaurants, $4,000 getaways to Nantucket, and Center Court tickets at Wimbledon. But Smith is still satisfied. "These events are incredibly difficult," she says. "I learned that I can only do what I can do." She gets home at 1 A.M. and returns the next morning to help clean up. She plans on holding another Bling Bling the following year.

Faustman and her group still have to make it back to Kansas City; as the car rolls through the Missouri cornfields in the dead of night, she falls asleep.

While $11 million would fund the first phase of the trial, Faustman figures she would need $30 million to complete all three phases. The most obvious contributor would be the JDRF, which in 2005 allocated $98 million to research and was in the midst of a five-year, $1 billion fundraising campaign. Faustman applied to the group for grants and had good reason to believe her requests would be well received. She had been a member of the JDRF's Scientific Advisory Board for five years and was its chair for two. The JDRF had given her money for previous research, and some of the board members held her in high regard. Gail Pressberg, the chairperson of JDRF's program development, said, "The respect for Dr. Faustman was such that we pulled together a workshop for scientists" to discuss regenerative therapies, which occurred in March 2004. The JDRF publicized Faustman's regeneration work in its "Research E-Newsletter No. 36," saying it raised "the possibility of new approaches to replacing diseased organs and tissues." Robert Goldstein, the JDRF's chief scientific officer, was asked about Faustman's research in a Senate hearing on adult stem cells, and he said it was "terrific. It is proof of something in animals that needs to be translated to people."*

Finally, the principal investigator for the trial would be Faustman's colleague David Nathan,[†] who played critical roles in two landmark diabetes studies — the Diabetes Control and Complications Trial and the Diabetes Prevention Program (which demonstrated that lifestyle changes can prevent type 2). Nathan, according to Francine Kaufman, the former president of the ADA, "is one of the most gifted clinical researchers in the field of diabetes," and his role would presumably give the JDRF requests added weight.

But the JDRF rejected three applications to fund the trial. The first request, in 2002, was for $10 million; the last, rejected on June 17, 2004, was for $660,000 a year for three years. The denials outraged Faustman's supporters, who couldn't believe that an organization dedicated to finding a cure would not fund a government-approved trial, and they vented their anger through e-mails and phone calls — ten to twenty a week, for months — demanding an explanation. Some supporters accused the JDRF of holding a vendetta against Faustman or favoring scientists who are associated with the JDRF. "It was tough to get any work done," said Richard Insel, the JDRF's executive vice president of research.

---

* The JDRF later said that Goldstein was talking about regeneration work in general.
† Nathan is the author's physician.

Faustman fueled the anger by saying she had been victimized by the rival scientists who judged her applications. The peer review process, used by the JDRF, NIH, and other institutions, assumes that only fellow scientists are qualified to evaluate proposed experiments, but Faustman says the process stifles innovation.* "In what other segment of the economy do your direct competitors get to approve your ideas?" she asks. "If I'm running a hotel chain, and every time I want to open a new hotel in a new city I have to ask Marriott's permission and Hilton's permission and Days Inn's permission, what kind of hotel is going to get funded? They'll say, 'You can open a hotel in the same city with the same structure, the same beds, the same everything. But a new hotel with a new format? No way.'" Defenders of peer review say that scientists can put aside ego, ambition, and greed to select the most promising applications. Faustman says that's absurd: "Human nature is human nature."

Faustman and Avruch also tried raising money from pharmaceutical companies but were turned away when — according to Avruch — they realized the proposed cure didn't involve a product they could develop. The first leg of the trial would require subjects taking an off-patent drug, Bacillus Calmette-Guerin (BCG), which costs only $11 a vial, and the specter of a trial for a low-cost cure that couldn't get funded stirred up conspiracy theories. These have long existed among diabetics. The disease, after all, has benefited researchers and doctors while enriching drug company executives, employees, and shareholders. Some patients idly speculate that the cure exists in the vaults of Eli Lilly or is commonly known among companies that make insulin. Other diabetics disavow conspiracies in general ("I don't believe in the grassy knoll theory") but insist that government and corporate leaders will never fight to eradicate a disease that now generates billions in profit every year. "There's too much money," goes the typical comment, "for them to ever find a cure."

Faustman's failure to secure funding led to some regrettable statements by her supporters, who claimed that the JDRF was more interested in protecting the salaries of its top executives than in finding a cure. Alternatively, they said, the JDRF wanted to protect its considerable investments in islet cell research against a proposed cure that

---

* Lay reviewers also evaluate JDRF applications.

would render those transplants obsolete. (Given a boost after the Edmonton protocol, the transplants now take place at seventy-eight centers worldwide.) "It's plain and simple," says Sue Root, one of Faustman's most active supporters. "Dr. Faustman's research is threatening. It's a cure without the need of pancreas transplants, islet transplants, embryonic stem cells, or expensive immune suppression drugs — everything science has been working on for the last twenty-plus years."

Referring to islet cell researchers, Faustman says, "I say publicly, for me to be right, other people don't have to be wrong. But everybody knows, if I'm right, they're out of business."

The JDRF said it could not disclose why it refused to fund Faustman because its rejection notices are confidential. It could say only that her applications failed to meet the standards of its science and lay review committees. The controversy escalated on November 9, 2004, when the *New York Times* published a story about Faustman and the difficulties "that creative scientists can have when their work questions conventional wisdom and runs into entrenched interests." The article noted that she had cured diabetes in mice, "something no one else had done," but she could find no financial backers for the next phase of her research.

The story also quoted Anita Chong, a researcher at the University of Chicago, who received funding from the JDRF to try to replicate Faustman's work. She said that while her research is still under way, "so far, what we have done replicates what [Faustman] has done." Two days later, she sent an e-mail to Faustman, saying, "I am only starting to realize the conflicts that underlie the politics and funding in diabetes research. I received a number of phone calls after the article came out, in particular from the folks at JDRF. They were very concerned that my quote could be construed as meaning that I have been able to replicate your work in its entirety."

The day after the article was published, two of Faustman's colleagues at Harvard sent a letter to the *Times,* challenging "some of the more glaring of the many incorrect statements" in the story that were favorable to Faustman. The letter was signed by a high-powered husband-and-wife team in immunology — Diane Mathis and Christophe Benoist. Both members of the prestigious National Academy of Sciences, they are co-heads of the Section on Immunology and Immunogenetics at the Joslin Diabetes Center and also heads of the JDRF Center for Immunological Tolerance in Type 1 Diabetes at Harvard. Mathis, in particular, is one of the most highly respected scientists in the field.

While their letter asserted numerous reporting errors, the *Times,* after receiving a lengthy response from Faustman, published no corrections or any portion of the letter. In the insular world of immunology, within the overlapping agendas of Mass. General, Harvard, and Joslin, the dustup might have been easily dismissed as an intramural squabble, but this letter had legs. The JDRF distributed it to its chapters around the country, which Faustman rightfully believed hurt her fundraising. Most donors who want to contribute toward a cure have little opportunity to make independent judgments about researchers. They call the experts, and the organization whose very purpose is to cure diabetes — and would presumably have the best experts — is the JDRF. Now that organization gave potential donors a disturbing view of Faustman's work.

According to Mathis and Benoist, Faustman's claim, reported by the *Times,* that she is the only scientist to have cured diabetic mice is "patently false," and they cite a paper published in 1994 that described another experiment that cured the animal.

But Faustman's position was that she was the first to cure mice with end-stage autoimmune diabetes without using antirejection drugs. The 1994 study, published in the *Proceedings of the National Academy of Sciences,* involved mice that had only been diabetic for seven days and were treated with immunosuppressants — failing to achieve, according to Faustman, "a common-sense standard of what constitutes a cure."

Mathis and Benoist also raised the possibility that Faustman's trial could harm its subjects, for it would require "injecting . . . an enormous quantity of spleen cells — the equivalent in humans would be the injection of cells from five human spleens over a period of four years. This procedure simply can't be done in humans." They cite no references on when or where Faustman made such a proposal. They later state: "To protect patients from useless or even harmful interventions, any proposed clinical trial must pass scientific (and sometimes lay) review before being funded." In an interview with JDRF officials, Richard Insel cites the spleen cell issue as well: "The problem is, we can't do spleen cell transplants."

A bigger problem might be this: her critics cite no evidence that Faustman has ever made such a proposal. Faustman did inject spleen cells into mice, but she says she has never proposed injecting them into people. "There is no human protocol," she says. "We've been attacked for something we have not proposed."

Faustman's final rejection letter from the JDRF, which she shared with me, made no reference to spleen cells. While she has written and spoken at length about her research, I could find no record of her proposing to inject spleen cells into humans. William Ahearn, JDRF's vice president of strategic communications and information technology, says: "It was public up until the time of the *New York Times* article." I asked that he show me the records. He never did.

Responding by e-mail, Mathis and Benoist concede that Faustman has never proposed injecting spleen cells into humans. Instead, they write, "we merely cited the protocol that was published in her science paper . . . which was impractical/impossible in humans because of the massive transfer of splenocytes it would have required." Apparently, it never occurred to Mathis and Benoist to simply ask their Harvard colleague what she proposed to do.

Their letter to the *Times* closed with a self-serving plea: "We offer our most heartfelt apologies, on behalf of Dr. Faustman . . . for having [diabetics'] expectations cruelly raised." But "there are other avenues that truly do offer hope."

The JDRF's decision to distribute the letter was like "a cornered animal lashing out," Debra Hull says. It was one thing to reject Faustman's request for a grant but something else entirely to try to discredit her. Merrill Goozner, director of the Integrity in Science Project for the Center for Science in the Public Interest, criticized Mathis and Benoist, telling *Diabetes Health:* "It is shocking to see that scientists, rather than evaluating something on its merits, would spend so much time attacking the messenger. You have to wonder, what is their real motivation? You would think that scientists connected with the JDRF would be pursuing every effective cure, not attacking approaches that rival their own."

JDRF officials say they distributed the letter because confidentiality agreements prevented their responding to the many questions and attacks from Faustman's supporters. The letter, Ahearn says, "was a proxy for our ability to say, well, look, there might be two sides to this story."

Faustman's critics give her supporters little credit. In an interview with *National Journal,* Mathis said that Faustman has "created a cult following" and that the researchers "at the JDRF have a much more scientific view of things than the parents." According to Insel, Faustman's popularity "says to me if you — or her people — go out there and say you can cure something without any kind of scientific underpinning and validation, you will recruit an army of vocal people to respond." Others are more blunt, if less public, in their description of her fans: "They're irrational."

The claim has a familiar ring. It was said by the ADA's doctors of the rebellious parents who founded the JDRF, but now the JDRF is the medical establishment fending off the insurgents. There is some truth to the criticism of Faustman. In her public presentations, she doesn't convey the enormous difficulty in translating any finding from mice to men — mice are laboratory animals that live in germ-free conditions and can be easily manipulated — and she does oversimplify a complex undertaking, leaving the misimpression that her trial, if successful, would cure diabetes. In fact, she has approval to conduct only the first part of a two-leg trial — all of which she spells out in newsletters and on the Iacocca Web site but not in public presentations.

On the other hand, the JDRF and its scientists engage in rank hypocrisy when they accuse Faustman of "cruelly" raising the hopes of patients, for that has been the JDRF's template since its inception. At the time of the Faustman controversy, the organization was running an ad campaign featuring a precious boy on his father's shoulders, the youngster reaching up toward a basketball net, the ball resting on the lip of the rim. The message read: "This close to a cure." It's poignant. It's powerful. And it's nonsense. We are nowhere close to a cure, but the campaign no doubt generated contributions.

The *Times* article also prompted a letter to the newspaper from Joslin's Ronald Kahn, who was more measured in his criticism but said the story contained "many inaccurate statements" and presented "an overly optimistic view" of Faustman's work. Ironically, Kahn treated Mary Iacocca and is the Mary K. Iacocca Professor of Medicine at Harvard. Lee Iacocca says he has sent letters to Kahn about Faustman, to no avail, and he isn't worried that the JDRF, Joslin, and two of the country's leading immunologists oppose his researcher. "When they're that fearful, she must be on to something," he says.

Unknown to most of Faustman's supporters, the response to her regeneration work reflected her critics' view that her previous research was unsound, contending that her experiments in the early 1990s could not be reproduced. The JDRF's rejection letter said Faustman's "line of investigation should be continued" but apparently not by Faustman, alleging that she had a history of publishing results "that do not accord with the laboratory of others." It's a serious charge. Reproducibility is a scientific cornerstone; it allows other researchers to build upon findings as a mason constructs a tower. If the results cannot be reproduced, the whole thing collapses. At least two papers in peer-reviewed journals either have cited inconsistencies with Faustman's work or have described researchers' inability to repeat the experiments. Faustman, however, identifies shortcomings in those studies — they used fewer test subjects, for example, than she did — while pointing to other papers that she says have confirmed her work. Sifting through the claims and counterclaims is difficult. Avruch, at one point, was instructed by his supervisors to review her primary data on a particular experiment. When he did, they checked out. "Most of the things she has found turned out to be true," he says, though the controversy has hurt her chances for promotion. She remains an associate professor at Harvard.

While scientists should be challenged, the challenges should come through lab experiments that lead to scientific clarity. But that doesn't always happen with Faustman. In one instance, she was giving a talk to students at the medical school when a strange man in the back of the room began arguing with her. Faustman had no idea who it was but later said, "He had steam coming out of his ears." Her accuser was Christophe Benoist, who apparently believed that the classroom was a fitting place to confront a rival. Some attacks are highly personal. For example, when the Autoimmune Disease Research Foundation, a nonprofit organization, tried to raise money for Faustman, it received an e-mail from David Serreze, a senior staff scientist at the Jackson Laboratory in Bar Harbor, Maine. Focusing on the immune system and diabetes, and thus a competitor of Faustman's, he wrote to the foundation that he knew well the "far-fetched Faustmanic claims," for she had engaged in "vapor science" and "scientific fraud," adding, "I feel it is important that you have this information since I also know how personally charming Faustman can be to lay people when selling her bill of goods (ever see *The Music Man*)." Serreze also wrote, "If she had any NIH funding, I and others would be requesting a federal investigation of scientific misconduct." Actually, Faustman has received twenty grants from the NIH, according to its Web site.

Faustman describes Serreze as "mentally disturbed. Normal people don't write letters like that." She threatened to sue him for defamation of character, which had one immediate effect: Serreze sits on the JDRF Scientific Review Committee, which evaluated Faustman's applications. On advice from counsel, he recused himself from grading his peer's request.

In an interview, Serreze says he can no longer talk about Faustman, though he did acknowledge that the e-mail "may have been too strong." He sent me a far more professional critique of Faustman's work, citing articles that he says support his position that her work has not been reproduced. Our conversation covered a wide range of issues in diabetes, and Serreze impressed me as thoughtful and conscientious, certainly not "mentally disturbed," which made his unseemly attack on Faustman all the more regrettable.

Faustman has made mistakes that have confirmed the perception that she's shallow or frivolous. For example, Faustman gave four dead diabetic mice that she had cured to Larry Raff, the founder of the Autoimmune Disease Research Foundation. According to Faustman, Raff told her he was going to stuff the mice and give them to a major donor to Mass. General; she then confirmed with the hospital's legal department that such a gift did not violate any rules. But Raff then announced he was going to hold an auction for the mice, placing each animal in its own jewel box. The proceeds would go to

Faustman's research. The idea, of course, was ghastly — animal research is not supposed to be used for fundraising gimmicks — and when word of the auction spread, Mass. General squashed the idea, but not before the mice appeared on eBay.

Faustman says she never approved the auction, and she eventually severed ties with Raff's organization. Raff disputes her account. "She didn't stop us from doing it, and she provided the mice," he says.

Raff acknowledges that putting the animals on eBay was his idea alone, but other scientists blame Faustman. "People who already hated her took this as evidence that she's psycho, and those who were responsible for her and the reputation of Mass. General were deeply offended, and so was I," Avruch says. "As I've learned over the years, you have to keep an eye on her. She doesn't have any gift for self-protection. She's just out there."

Nonetheless, he bristles at the ad hominem attacks, saying they partly reflect the airy style of her presentations. More important, he maintains that she poses a threat to an industry that values me-too science over innovation and risk-taking.

He is particularly offended by her critics at Joslin. "There are two things working there," he says. "One is fundamental self-righteousness. 'This can't be right. I don't like her. She's wrong.' The other is self-protection of resources. A lot of money goes in there based on the perception that they're doing cutting-edge research and they're moving toward a cure." Avruch also met with Mathis and Benoist, whom he calls "those assassins." "What it comes down to is this," he says. "She offends them tremendously. The idea that she is getting recognition offends them tremendously and makes them sufficiently crazy to act in this personally vicious way." Asked about the JDRF's response to Faustman, Avruch says: "They're worms."

For Denise Faustman, the headlines on March 24, 2006, could not have been better. Her photograph was above the *Wall Street Journal's* front-page banner with the blurb: "Confirming Hope on Diabetes." The *New York Times* also published her photograph, on A11, with the headline: "A Controversial Therapy for Diabetes Is Verified." The stories summarized three new papers in *Science,* in which researchers tried to reproduce Faustman's regeneration experiments. Conducting the studies, in addition to the University of Chicago, were Washington University and Joslin; the lead investigators at Joslin were Diane Mathis and Christophe Benoist.

The buoyant headlines were understandable. Each group had cured the diabetic mice, though at lower success rates than Faustman, and each group reported that the animals had beta cells in their pancreas. "Our studies confirm that autoimmune diabetes can be reversed and that sufficient endogenous β cell mass can be restored to cure diabetic NOD mice with the treatment protocol developed by Faustman and colleagues," wrote Louis Philipson, the senior author of the University of Chicago group.

Faustman hailed the results as "fantastic," confirming that mice can be cured through regeneration. "That's the beauty of it," she told the *Journal*. "The animals' own pancreas did this."

But the results were far short of complete vindication. All three studies failed to confirm Faustman's finding that the spleen was a source of new beta cells, undermining the possibility that the organ has some curative potential. While Faustman remains confident that her results were correct, the inability of others to reproduce them renewed concerns that her work is sloppy or unreliable.

Moreover, the compound used to cure the mice, CFA, can't be used in humans, and the compound that Faustman wants to use — BCG — has had mixed results in previous diabetes trials. In their paper, Mathis and Benoist acknowledge the "autonomous efficacy [of CFA] demonstrated in our study," but say that the failure of BCG to produce positive outcomes in three previous studies should negate its treatment on "diabetic humans." There is even dispute whether beta cells truly regenerated — in which case new cells were created — or whether residual islets simply "recovered," suggesting they are too weak to cure diabetes permanently.

Nonetheless, Faustman is credited for shattering the conventional wisdom that curing diabetes, even in mice, requires a cellular transplant after the immune system has been tamed. As Jeffrey Bluestone, director of the Diabetes Center and the Immune Tolerance Network at the University of California, San Francisco, told the *Times*, "What these papers suggest is that . . . even a long time after the mice are diabetic, they still have the capacity to develop new beta cells. That's very exciting."

Even before the *Science* papers were published, Faustman had secured the money to conduct her clinical trials. Though long retired from the company, Iacocca persuaded Chrysler to become an official sponsor of Faustman's research, including a pledge from Chrysler, Jeep, and Dodge dealers to donate one dollar for every car sold from November 1, 2005, until the end of 2006. It was part of a negotiation that brought back Iacocca to promote Chrysler cars in television commercials. Faustman hopes to begin the trial by 2007.

If she fails to cure diabetes, her critics will claim vindication, but at least she is willing to stake her reputation on the long shot. That distinguishes her from those scientists who content themselves with safe, predictable studies that produce grants but, in the ocean of medical research, barely create a ripple. In an industry constantly pushing halfway technologies with the false promise of diabetic utopia, Faustman has given her supporters reason for hope, and they will appreciate her sacrifice even if she fails.

Faustman herself recognizes the precarious nature of her work. "It'd be much nicer for my career if I stayed working in the mouse," she says. "It's a much more comfortable existence. You get invited to more cocktail parties, and you're loved by more people. Nobody would care what I did. It only becomes a problem when I say, okay, let's try humans."

CHAPTER 13

The Price of Survival

SOME THINGS WORK.

We know what doesn't work in diabetes care — the traditional approach of physicians telling patients what to do. But other models have had more success, often under the rubric of "disease management" for chronic ailments. The most important provider for any chronic disease is the patient himself, which places the highest priority on education and training. This need is problematic, not just because insurers reimburse education poorly. Some health care experts suspect that the majority of diabetics will never have the resources or resolve to care for themselves properly. Behavior modification, they argue, is too elusive and too difficult to translate to the masses. But I believe most diabetics, given the tools and training, are willing to discipline themselves to stay healthy. As one sixty-year-old patient e-mailed my brother: "Dr. Hirsch: I do have retinopathy. I know what you would say: Let's use this as a motivation. I want to do that. How do we proceed to kick my ass?"

Consider the success of HealthPartners, a health maintenance organization in Minneapolis, which began a diabetes management program in 1992. At the time, HealthPartners helped fund the Institute for Clinical Systems Improvement (ICSI), which advises medical organizations in Minnesota. HealthPartners itself has 3,700 primary care physicians and 4,500 specialists, and with ICSI's assistance, its diabetes program provides clinical guidelines and goals for doctors. While those goals may be well known to diabetologists and most endocrinologists, they are not familiar to most general practitioners. HealthPartners also funds educational mailings for patients and, using the results of A1c and cholesterol tests, generates a list of those patients at risk. They are then contacted by special "diabetes resource nurses" from HealthPartners' clinics for additional education and self-management support. The HMO also distributes a "diabetes performance profile" to each physician, which compares the patient's lab results with the average results in the clinic and the medical group. HealthPartners uses financial incentives as well. It pays between $75,000 and $250,000 to medical groups that hit "stretch targets" in diabetes care. Though the physicians say the bonuses are not large enough to change their treatment, the money helps support the program's added administrative costs.

The initiative's impact has been significant. From 1994 to 2000, the mean A1c fell from 8.7 to 7.7 percent. (The original number of patients, 6,292, was reduced to 3,535 due to attrition.) Average LDL cholesterol levels also declined, from 130 to less than 100mg/dL.

These kinds of programs, and there have been several, simply confirm the results of the Diabetes Control and Complications Trial: with proper support, patients will achieve much better outcomes. The real question is whether these programs make sense financially. David M. Cutler, a professor of economics at Harvard University, investigated the HealthPartners program. In the first few years, he calculated, the program increased HealthPartners' costs, reflecting high start-up expenses, including staffing and computer systems. What makes diabetes so expensive is the treatment of complications — a lower extremity amputation, for example, costs between $30,000 and $45,000 — so the financial benefits of improved care are not realized until the number of complications is reduced. "It is therefore unlikely," Cutler concluded, "that a program can break even before eight to ten years."

Few insurers or employers have the luxury of waiting that long to realize a profit, particularly if the patients themselves switch jobs and move to other plans, thereby negating any benefit. On the other hand, Cutler estimated the "societal benefits" of better care, including improved quality of life and increased worker productivity, and concluded that while HealthPartners' program cost about $330 per patient, the "net societal benefit" was $30,000 per patient. "At the societal level, comprehensive disease management programs are clearly worth the investment," Cutler said.

There are obvious policy implications. Government insurance programs, particularly Medicare, have an incentive to support diabetes programs, because the reduction in costs from complications will likely occur when the patient is older and enrolled in Medicare. "It seems reasonable to ask whether Medicare could be charged some amount to subsidize disease management programs," Cutler said.

In fact, Medicare must be the catalyst for reform, because its payment methods are often copied by private insurers and because its size can influence the behavior of providers. As a *JAMA* article on chronic care notes, "If Medicare paid for chronic care start-up costs (including information systems), reimbursed nonphysician personnel who provide chronic care services, and increased reimbursement rates for provider organizations with superior performance, arguing the business case for chronic care would be easier."

Other studies have demonstrated that improved diabetes care reduces costs in the short term as well. The Group Health Cooperative of Puget Sound, for example, found that its intensively managed diabetics, who achieved lower A1c's, produced a savings of $685 to $950 per patient annually compared to conventionally treated diabetics. The savings reflected fewer hospital admissions, emergency room visits, and physician consultations. What's clear in many of these studies is the need to shift the burden of care from physicians to equally effective but less costly nurse specialists, nutritionists, and diabetes educators. Medical assistants, for example, can be trained to look for lesions on feet and to check sensation with microfilament.

Pharmacists are perhaps our most underused resource; diabetics have more encounters with them than with their physicians. A study by Kaiser Permanente, for example, showed that patients had better outcomes at lower costs when their pharmacists gave them education materials, monitored their progress, and offered feedback to doctors. A good pharmacist will alert diabetics to new drugs while warning them about the potential side effects of combining existing ones — a critical point for the many type 2 patients who require a smorgasbord of medications for various illnesses. Some pharmacists even inspect patients' feet. The problem, of course, is that few pharmacists have time for individual attention beyond the perfunctory "Do you have any questions for the pharmacist?" At my CVS drugstore, the line for prescriptions is sometimes ten people deep, so by the time I'm at the counter, I have little interest in inquiring about new therapies. Nonetheless, after Garrett was diagnosed, I did make the effort to get to know one of the pharmacists so we're on a first-name basis, and I know he'll respond if I call with an emergency.

Other studies have shown that it's even possible to improve outcomes in the hardest to reach low-income and minority populations. One study at three clinics in California involved 362 type 2 Medicaid patients, 55 percent of whom were minorities. The patients who were intensively treated received assistance from registered nurses and dieticians, collaborating with an endocrinologist. Each patient received a glucose meter and individual instruction on how to use it as well as lessons on diet and exercise. Transportation was provided for medical appointments, and missed appointments were rescheduled. Ancillary medical services, such as eye and foot care, were provided. Over a two-year period, the intensively managed patients improved their A1c's from an average of 9.54 to 7.66 percent. The study's coordinator, Lois Jovanovic, also tried to remove cultural barriers. Many Mexican-American patients, for example, feared insulin because they recalled that it had been given to an aging diabetic ancestor long after it could do much good. Insulin therefore seemed like a precursor of death. "The word is terrifying, because it's the drug that killed everyone in the family," says Jovanovic, who is the director of research and chief scientific officer of Sansum Diabetes Research Institute in Santa Barbara, California. To convince Mexican-American patients to accept insulin therapy, she didn't say its name. She used insulin pens, calling them *las plumas.* There were *plumas* with short-acting clear insulin, which she called *clara,* and long-acting cloudy fluid, or *nublada.* "You can teach insulin therapy very quickly with pens," Jovanovic says. "Just tell them it's medicine."

One of the most important movements in health care reform, and one with a direct impact on diabetes, is pay-for-performance, or giving bonuses to providers whose patients achieve good results. I believe it's a bad idea whose time has come.

It's a bad idea because it rewards doctors for doing what they should be obliged, professionally and morally, to do anyway: provide excellent care for their patients. It's also bad because many initiatives are tied to quantifiable outcomes — in diabetics, for example, A1c scores — so these programs could cause doctors to pick the healthiest patients while avoiding those at high risk. Low-income and minority diabetics, for example, tend to have higher A1c scores and increased rates of complications than affluent white diabetics, so the patients most in need could lose out in performance-driven reimbursement systems. Measuring care is also problematic because patients with the same condition may respond to the same treatment differently, raising the possibility that doctors will be rewarded or punished for results that are beyond their control.

Nevertheless, it's clear that many providers will never offer the care that diabetics need without significant financial incentives. The current system has failed because diabetes is a disease of details, and expecting providers to sort through and interpret those details, let alone educate patients on their meaning, is unrealistic under the current reimbursement rules. Incentives could also be used to ensure that patients get the basic tests — for glucose and cholesterol levels, eyes, nerves, and kidneys — that are needed to assess their health.

As Cutler writes about patients generally in *Your Money or Your Life,* this approach would force providers to change their entire orientation to care:

> If there were additional income to be earned by making sure cholesterol tests were performed, doctors would figure out how to increase testing. Physician offices would set up reminder systems. Nurses would contact patients to arrange convenient times. Outreach would be facilitated. Hospitals would similarly search for ways to improve bypass surgery mortality and cut postoperative infection . . . Health-based payments reward high-value services regardless of their intensity.

The momentum for pay-for-performance is building. Medicare, for example, has introduced pilot programs that pay hospitals and doctors bonuses for achieving better results, including improved blood sugar control for diabetics. Under the doctors' program, which covers about 200,000 patients, physicians will get bonuses if they measurably improve care for people with chronic diseases. They must also provide preventive services that will save Medicare money by keeping patients out of the hospital and eliminating unnecessary procedures. A similar program evaluates thirty-four performance measures of 280 hospitals, and Medicare is expected to pay $21 million in bonuses to the high-achieving hospitals over three years.

Medicare is one of many entities experimenting with pay-for-performance. About one hundred such programs have been created by insurers and employers, according to Med-Vantage, a consulting group. The largest is called Bridges to Excellence, created by a coalition of big companies, including General Electric, Ford Motor, Verizon, and Procter & Gamble. Their goal in part is to improve health outcomes by realigning incentives, and the organization identified diabetes, a costly, labor-intensive disease, as an ideal case study to determine if payments actually help. Based on actuarial analysis, the group concluded that diabetic patients who adhere to demonstrably healthy guidelines will save their companies or insurers $350 a year, primarily through reduced hospital stays. That figure became the basis of several pilot programs around the country. The organization initially proposed that for each diabetic employed by a Bridges company, qualifying physicians would receive half the savings, or $175 per patient per year. The doctors, however, said the program would work only if the patients were motivated, so they volunteered to redirect $75 of their payment to the diabetics themselves for medical supplies or participation in a weight-loss program.

Robert Galvin, the director of global health care at G.E., told me it's too early to determine the success of the diabetes program, which began in 2003 in Cincinnati and Louisville and later expanded to Massachusetts. Recruiting physicians, he acknowledges, has been a challenge. While some welcome an alternative approach, he says, "others resist change and are pretty suspicious when a payer comes to them and talks about quality."

Advocates of pay-for-performance emphasize that it's not about punishing bad doctors but rewarding good ones, and they believe other concerns can be addressed. High-risk patients, for example, can be evaluated with different criteria. What makes diabetes different from most diseases is that patients themselves play such an indispensable role in their own care, so a true rewards system would need to benefit them as well. Bridges to Excellence has the right idea in paying patients for successful outcomes, but the actual amount, $75, seems insufficient to change behavior. Galvin agrees. "Seventy-five dollars is not the answer," he says, "but it helps more than it hurts."

The clearest example of how properly aligned incentives can improve diabetes care can be found at the Department of Veterans Affairs (VA). Though once synonymous with poor treatment, the VA, under new leadership in the 1990s, began emphasizing disease management, prevention, and incentives for performance. Central to this effort was a huge investment in computers and software, upgrading existing systems and giving its physicians integrated electronic medical records on every patient. The absence of electronic records is one of the industry's most conspicuous shortcomings, but it particularly hurts a data-driven condition like diabetes, whose patients, suffering various maladies, might have paper records scattered in a half-dozen different places. But for any one of the VA's 800,000 diabetics who enters a VA facility, doctors have instant electronic access to an entire medical profile, including previous lab results, histories, performance goals, and treatment guidelines. The VA estimates that it costs about $78 per patient per year to operate its electronic record, roughly the equivalent of one blood test. The Institute of Medicine said the VA's "integrated health information system, including its framework for using performance measures to improve quality, is considered one of the best in the nation."

The VA's diabetic care has also been superior. In one study, 1,285 diabetics in the VA were compared to 6,929 diabetics from managed care sites. The VA patients more often received A1c testing, counseling, and eye and foot exams. In the VA group, 84 percent had A1cs under 8.5. In the managed care group, only 65 percent could make that claim. The VA patients also had better lipid control.

Technology has helped the agency improve care for all of its patients at its 1,200 sites. Since 1995, the VA says it has doubled the number of patients treated to 5.2 million while trimming its staff by about 12,000 people and cutting costs per patient by half. In 2003, the *New England Journal of Medicine* published a study that compared the VA's health facilities with Medicare on eleven measures of quality. On each count, the quality at the VA facilities was "significantly better." *Washington Monthly* describes veterans' hospitals as "the best care anywhere."

The VA's success is no mystery. It has an economic incentive to keep its patients healthy. When a veteran is hospitalized or requires special tests or procedures, the VA receives no additional revenue. It effectively loses money because the treatment is covered by the fixed appropriation it receives from Congress — $21 billion in 2005. Because the VA usually has a permanent relationship with its patients, it can justify costly computer upgrades or expensive disease management programs, for it will reap the financial rewards of healthier clients in years to come. For the rest of the health system — providers, insurers, employers — those upfront costs are seen as prohibitive. At the same time, the VA has implemented pay-for-performance incentives for its top executives and for individual facilities, which increases short-term costs in exchange for improved long-term outcomes. The system has shown that it can pursue quality without threatening its own financial viability. "If we've proven anything . . . in the last ten years, it is that quality is less expensive," said Jonathan B. Perlin, the VA's acting undersecretary for health.

Some health care reformers cite the VA's experience as a model for a single-payer system, though any proposals to move toward that structure would meet stiff resistance. The affluent and the powerful are well served by the current system; medical organizations, fearing a loss of control or income, typically resist any government intrusion; and single-payer approaches stir fears of health care rationing and shortages.

Even if financial incentives are provided, doctors will still fall short if they don't know what they're doing. My brother, for one, believes that is often the case, and he has the experience, as well as the credentials, to make such a claim. In 2004, the ADA named Irl the Physician Clinician of the Year, and in 2006 he won the Distinguished Endocrinologist award from the American Association of Clinical Endocrinologists. As a diabetic himself for more than four decades, Irl also stands as one of the community's leading role models.

Noting that most diabetics are treated by primary care physicians, he says that substandard care is the byproduct of inadequate training. Medical schools provide only one afternoon of diabetes training, excluding electives, in a four-year education. Forget the highly sophisticated management tools or the latest research on glycemic variability. When students or residents assist Irl, he must show them an insulin syringe, an insulin pen, and vials of insulin. "Most don't know what these look like," he says. "Even the fellows are clueless."

The best residency programs for diabetes care tend to be obstetrics and gynecology, because the attending physicians as well as the patients recognize the imperative of tight control during pregnancy. The problem with other residencies is that "most attendings practice diabetes the way they were taught when they were in residency," Irl says, adding that some programs don't even teach glucose monitoring. That might account for why so many diabetics in this country all but ignore testing their blood sugar.

Irl says that even if endocrinology attracts more people, it won't necessarily improve treatment, in part because government funds research more generously than clinical care. "Half of the people who go into endocrinology don't want to see diabetes," he says. "They stay in the labs and become rat doctors."

Irl proposes a special training course for primary care physicians who have completed their three years of training and are willing to emphasize diabetes care in underserved communities. The government should identify a half-dozen "centers of excellence" — places like Joslin, the Barbara Davis Center in Denver, and the Diabetes Research Institute in Miami — and pay doctors to train there for an additional six months to a year. While these physicians would not be board certified in endocrinology, they would get some kind of diabetes certificate — Irl says he'll make it himself if he has to — indicating expertise in the field. This would appeal to at least some general practitioners, Irl says, because they know they'll be seeing diabetics anyway; now they'll be better equipped. The government, of course, would have to foot the bill, but the money would be well spent. The epidemic will never come under control unless providers can give patients the tools they need. "It doesn't matter if you have a motivated patient who's willing to do everything," Irl says. "If he doesn't have the information, it won't do any good."

Why haven't medical school administrators and public officials more aggressively confronted the epidemic? Bureaucratic inertia is one reason, but diabetics have themselves to blame in part. The disorder is often called "the silent disease" because it can be so easily hidden. But it's also silent because relatively few patients, particularly type 2's, speak out. They are the true "silent majority." For example, in the late 1990s, Congress created the Diabetes Research Working Group to develop a road map for eliminating the disease. The group invited any individual or group to offer public testimony; ten applied for permission. Of the ten, nine represented type 1 patients, and one represented dental health in diabetes. Not a single person or group stepped forward to speak for type 2 diabetes. "I was shocked — flabbergasted," said Ronald Kahn, Joslin's president, who chaired the group. "I couldn't believe that no one came to speak, even though the epidemic is largely type 2."

In some ways, this reticence reflects the legacy of Elizabeth Evans Hughes — the shame and embarrassment that diabetics feel about their condition. But it's more understandable that a child or teenager or even a young adult would want to conceal the disease. Insecurity already runs high in those groups. Overweight type 2s are also reluctant to come forward if they believe they themselves are responsible for their condition, and minority patients — who already feel disenfranchised — are less inclined to advocate for their own health. What's odd about type 2 is the number of physically fit, accomplished professionals who still feel compelled to hide their condition.

Richard A. Smith, for example, was sixty years old when he was diagnosed with type 2. At the time, he was president and chairman of three large companies — Harcourt General, the Neiman Marcus Group, and the GC Companies — and a prominent civic leader in Boston. He could have been a powerful advocate, but, he recalls, "I damn well wasn't going to let anyone know that I carried a disease called diabetes. Why? It didn't in any way inhibit my job, but it would inhibit the perception of what I could do. Therefore, it was something I was not going to talk about." Asked if he feared the reaction from investors, Smith said he was primarily concerned that his standing would fall among employees — an improbable scenario, given his power and stature.

Smith is now eighty and has retired from most of his corporate responsibilities. Several years ago, he joined the board of Joslin and de-

cided — as he says — to "come out of the closet," but he still believes
that the stigma against diabetes deters patients from advocating on
their own behalf. "I've known maybe fifty CEOs around Boston," he
says. "The incidence in the population would tell me that maybe ten
of them have diabetes. I know of none. That doesn't make it right or
wrong. That's just the way it is."

But it doesn't have to be. The JDRF, to its credit, has considerable
political strength, but in advocating for type 1's, it represents a small
fraction of the diabetic population. Politically speaking, the group as
a whole is a sleeping giant. Only after all patients, types 1 and 2, 20
million and counting, make their voices heard will they generate the
attention they deserve. Just as veterans demand funding for VA hospitals and gays for AIDS research, diabetics could — and should —
make similar demands, prodding the government to act aggressively
against the country's leading public health crisis.

In one respect, the disease is like every other. Whether it's for
better treatments or possible cures, it needs money. It would benefit
if it had the equivalent of a Howard Hughes, the industry mogul
whose institute has awarded more than $1 billion for biomedical re-
search since 1988. But even with such a patron, it would still need
what polio had: a president or a widely admired political leader, like a
Colin Powell or John McCain, who would personify the issue and
make it a national priority.

Newt Gingrich, as the Speaker of the House, demonstrated how
even one of the country's most divisive politicians could rally support for diabetes. His mother-in-law was diagnosed in her sixties, so
Gingrich spearheaded landmark legislation in 1997 that authorized a
$2.1 billion expansion of Medicare benefits to diabetics while also
boosting funding for diabetes research, screening, and treatment.
Since leaving office, Gingrich has become a leading advocate for re-
structuring the health system to promote preventive care, and his
name is revered in many diabetic circles. Surely other politicians, if
only for their own self-interest, would recognize the benefits of join-
ing this crusade.

I used to have a perverse dream in which researchers eliminated all
forms of diabetes through prevention, just as Sabine and Salk did for

polio. New diagnoses end; Nobels are awarded; dancers take to the street. Those of us with the disease soldier on — with a difference. No more jubilant ballroom fundraising galas, no more bikeathons or walkathons or telethons, no more carnival conventions for the medical companies' newest gadgets, no more catfights among scientists. The camps for diabetic children are closed, the ADA and the JDRF shut down. The Joslin Center is turned into a museum, with smiling escorts, interactive videos, and an infrared glucose meter that confirms the good health of all visitors. The drug and medical device companies turn their energy to more urgent diseases, though Eli Lilly continues to make insulin, for old time's sake.

One by one, the remaining diabetics die off. The older ones, of course, go first, until we are down to the last generation, its members appearing at county fairs, waving to appreciative audiences as triumphant proof that science has indeed conquered an implacable foe. Their numbers dwindle, with a Web site (www.Diabetics AreExtinct.com) counting down the demise of each one. Finally, one guy is left, perhaps a taxi driver in Wichita, and when he takes his last breath his passing is mentioned by all the newspapers, and he is given a proper burial, with a syringe and insulin bottle laid gently in his casket, and the disease is over.

That's my dream. Except it may be reality, for prevention is now seen as more feasible than a cure for either type 1 or type 2, and that is where a good deal of the research money is going.

In recent years, for example, the NIH funded two clinical trials that sought to prevent type 1 diabetes in at-risk patients. One trial used insulin injections; the other, oral insulin. Both failed. The NIH has now committed an estimated $151 million over seven years to fund TrialNet, whose goals are the prevention or reversal of disease in new patients who retain some beta cell function. Another clinical trial is studying the genes of diabetics as a possible avenue for prevention. Another will examine environmental triggers, such as infectious agents or diet, which potentially cause the disease. Yet another is comparing different baby formulas to see if they could cause it.

Recent clinical studies, in fact, have raised expectations that the disease can be reversed in new patients. Researchers targeted an immune system molecule known as CD3 with an antibody, and the intervention temporarily halted the progression of the disease. In one study, patients who received the antibody for six days after their diagnosis required less supplemental insulin to maintain normal blood sugars than patients who received a placebo, and this benefit continued to work eighteen months after treatment. The results suggested that residual beta cell function can be preserved by modulating the autoimmune attack, changing the course of the disease once it has begun and offering new strategies for prevention. "I predict it will be used for prevention, once proven safe," says Richard Insel of the JDRF, which has funded much of the CD3 research. The treatment would not help patients without beta cell reserves, but Insel says that scientists are going to combine the antibody with possible regeneration therapies. That the body can regenerate beta cells has not been proven, but it may well be diabetics' best hope: they're going to have to save themselves.

The challenge for type 2 diabetes is different, though no less daunting. For some patients, the cure is already known: a combination of weight loss, exercise, and diet will restore normal glycemia without the need for medication. At least 70 percent of type 2 patients would benefit from losing weight, but for all the millions poured into the weight-loss industry, obesity rates continue to rise. There are two possible solutions. One is medical — a new drug, for example, that changes the "energy balance" so that an extra 30 calories a day don't add another pound every few months or year. At least five obesity drugs are now in phase 2 or 3 trials. The other is broad lifestyle changes. As Francine Kaufman notes, we need school and community programs that promote better eating habits and more physical activity. That means eliminating soda and snack machines from schools; reversing the decline of physical education classes and athletic activities after school; building sidewalks, bike trails, and parks; increasing neighborhood safety so parents and children don't fear being outside; even encouraging the use of stairs instead of elevators. Contributions are required from parents, politicians, teachers, clergy, doctors, pharmacists, and land-use planners, to name just a few. Sweeping social transformations are hardly easy and probably not realistic, but they must be part of any campaign to slow the type

2 epidemic. And even those changes wouldn't affect the physically fit type 2 patients, whose disease is even more mysterious than that of the type 1 patient. Researchers recognize that type 1 involves an auto-immune trigger but know little more about type 2 other than that it involves various genetic defects.

It's also possible that the differences between the two categories have been exaggerated. According to a leader in the field, the cure for type 2 will be the same as the cure for type 1. "The defect that distin-guishes type 2 diabetes from simple obesity is the beta cell dysfunc-tion," says Jay Skyler at the University of Miami's DRI. "Therefore, when we cure type 1 diabetes — beta cell replacement therapy . . . whatever allows islet transplants to overcome the resource limitation and hopefully the immunological barrier — [that] will also be the cure for type 2."

If complacency has hurt the cause — at least in pressing for a cure — I plead guilty as well, though not without reason. For most of my dia-betic life, I never thought twice about a cure. Never. I would see occa-sional newspaper headlines or maybe overhear something on televi-sion or the radio, but I never actually read or followed any story on the subject. I've also known many smart doctors and researchers, including my brother, but I never asked any questions. I had good reason to be incurious. First, it was a form of self-protection. If I had begun investigating the subject, prowling medical journals for scraps of hope, I would only be setting myself up for disappointment. I knew that researchers had been promising a cure since at least the 1970s and frustrating their followers at every turn. Why set myself up for heartbreak?

What's more, hoping for a cure or believing that one was near could be unhealthy. If I believed one was "around the corner" or "five years away" or even possible, it would have been easy to lapse into bad habits. Why go through the daily demands, frustrations, and indignities of tight management when a medical miracle would soon deliver me from my burdens? Diabetes is too taxing, too unfor-giving, to hold out hope, and I think many others fall into that cate-gory. It's the ultimate paradox of the disease: if you have it, you have to live your life as if you'll never be cured.

My attitude changed when Garrett was diagnosed. I understood, for the first time, the fury and bitterness that accompany this disease, and why parents are far more active than patients. The axiom that "you stand where you sit" is never more true than with this disorder. Take advances in diabetes management. Based on over a thousand interviews, letters, and e-mails, I've concluded that if you don't have diabetes, you think the glass is half full. If you have it, you think it's half empty. And if your child has it, you think the damn thing's empty.

Perspectives also differ on the cure. As a teenager, I saw Walter Cronkite on CBS introduce a story about diabetes by saying that patients "have a fifty-fifty chance to reach age forty." I confronted my mortality at a young age, which wasn't all bad: I recognized the importance of self-management for survival. But Garrett's diagnosis forced me to confront every parent's worst nightmare — their own child's mortality. There is no consolation beyond hoping for a cure. If having the disease means you have to live your life as if there will be no cure, there is a corollary: if your child has it, you have to live your life as if the cure is around the corner.

Garrett's diagnosis caused me to look at the disease, literally, in a different light. One day, I walked out of a steamy shower, stood beneath a bedroom skylight, and happened to look at my fingertips; gray and wrinkled, they revealed the microscopic dots left by thousands of finger pricks. A year earlier, I would have been amused by this bizarre tableau, a road map of my past, but now it was a glimpse at Garrett's future. I didn't look again.

What did interest me was the cure — not just the likelihood of one but the likelihood of one *immediately.* I feared Garrett would experience his entire childhood under a cloud, mindful of the unfairness of life, aware of some distant loss. If he were cured in, say, five years, he would be eight, old enough to know sorrow, perhaps, but young enough to renew his innocence.

I had already planned to visit or interview top scientists in the field, but now my inquiries had less to do with the book than with my son. Why isn't there a cure? How far away are you? What, in God's name, have you people been doing for the past thirty years? I felt the same mix of anger and anticipation that has long created unrealistic expectations for curing this disease — a desire, Michael Bliss observes, that seems to be most common in the United States, reflecting "a complicated mix of American optimism, evangelicalism, and exceptionalism." Our penchant for declaring "wars" on cancer, malaria, AIDS, and other diseases, not to mention poverty, implies that such wars are winnable.

But as I quickly discovered, they aren't, at least not this one, not any time soon, and I was forced to recognize that Garrett had no chance of being cured during his childhood or well beyond. Indeed, I could not find a single clinical trial anywhere in the world to cure type 1 diabetes. (Denise Faustman hopes to begin one in 2007.) And even if a trial were to succeed, it would take at least ten years before any treatment would be available. While we can improve ways to maintain normal glycemia — from continuous glucose sensors to cellular transplants — we have not, and perhaps never will, resolve the underlying immunological defect: the disease itself. Will there ever be a cure? While there are certainly pockets of optimism, expectations are generally low. The reason is simple. The immune system will not easily allow any mortal to reeducate it, retrain it, or tame it.

I spoke to Richard Kahn, the ADA's scientific director. He said that if anything good came out of the AIDS epidemic, it was the "awakening force" of how complex the immune system is and how little we understand it.

I spoke to Ronald Arky, a former ADA president, a professor of medicine at Harvard, and the chief of diabetes and metabolism at the Brigham and Women's Hospital. He said, "The more we learn, the further we go from a generic cure."

I spoke to Maria Buse, the distinguished professor of medicine and biochemistry/molecular biology at the Medical University of South Carolina who's been researching diabetes for more than fifty years. "We can't change genes," she said. "There is no cure for a genetic defect unless you want to reengineer children." The time and money are better spent trying to modify behavior or developing better pharmacological therapies, according to Buse. "We would settle for optimal management." When I said that parents wouldn't accept that, she waved her hand. "You talk like a diabetic," she said.

You know you're a long way from a cure when the very word is being avoided. "Conquer" is in vogue. Mass. General Hospital publishes a brochure, "Conquering Diabetes: How Does It Begin?" The Diabetes Research Working Group produced a report, "Conquering Diabetes: A Strategic Plan for the Twenty-First Century." A highly regarded California doctor has written a new book called *Conquering Diabetes*.

Joslin's Ronald Kahn explained why the center's motto is "Conquering Diabetes in All Its Forms." The word "cure" conjures improbable expectations. "If I give you an antibiotic for pneumonia and the pneumonia completely goes away, then I've cured the pneumonia," he said. "But most of the things we talk about with type 1 diabetes don't make the intrinsic disease go away. In fact, the only way we'll make the intrinsic disease go away is more in the area of prevention than cure; that is, if we can find something that modulates the immune response and turns that autoimmune response off. Then we can say, 'Okay, this person was at genetic risk and we've really prevented the disease.' That we could call a true cure because he wouldn't get it."

The search for a cure is complicated by improved insulin therapies, because all "cures" come with side effects or tradeoffs. This is what separates diabetes from, say, kidney, heart, or liver disease. Transplants can save those patients, who are willing to accept immunosuppressant drugs because the alternative is worse. Let's assume, in diabetes, that regeneration does not occur. Even if scientists can modulate or retrain the immune system, patients would still need a cellular transplant to maintain normal blood sugars, and that means antirejection drugs. "If I want to cure you, I have to give you something that is better than the existing therapy," Kahn said, "and while the existing therapy is far from perfect, it allows most people to live good lives."

Dan Mintz, the researcher who began his work on islet cells in the 1970s, said he's appalled at the promiscuous use of the word "cure" by other scientists and organizations, having forbidden the term in his lab because it misled patients. He believes instead that patients and their families will support the research if they feel part of the process. "They understand, better than most, that the research scientist deals daily with failed experiments, unreasonable hypotheses, unproven notions, and sheer intuition as food for tomorrow's experiments," he said. Researchers' only promise should be that "they work harder, strive harder, and persist longer because they understand that it is someone's child who may be remedied."

I spoke with David Harlan, chief of the Islet and Autoimmunity Branch of the National Institute of Diabetes & Digestive & Kidney Diseases at the NIH. He is no more optimistic about a near-term cure than his colleagues, but he helped me make peace with a question that had long haunted me. Why does type 1 diabetes occur? It makes no sense, contradicting all of the body's survival mechanisms. Why would the very system that's supposed to defend the body be the architect of its destruction?

Harlan points out that there is a reason for diabetes, a good reason, one that should enhance our appreciation for the people who have it.

The reason requires a deeper understanding of the immune system. According to evolutionary immunologists, it consists of two broad categories of cells. The first, the innate system, is the more primitive and the first line of defense; it can recognize pathogens and draw on various weapons to eliminate them — phagocytic cells, for example, that consume them. Working closely with that primitive system is a more sophisticated one, the adaptive immune system, which first appeared hundreds of millions of years ago with the vertebrate evolution. What most people are familiar with — the system of T cells and B cells that remember how to fight specific germs the body has already been exposed to — evolved some 200 to 300 million years ago with the common predecessor for all mammal and bird species. The T cells and B cells have receptors on their surface that recognize only one antigen. B cells make antigen-specific antibodies; T cells are responsible for a "cell-mediated immune response," recognizing and destroying cells damaged by toxins or bacteria. These same T and B cells also direct the immune response against cells, tissue, or organs transplanted from another member of the same species.

The adult body has between 100 billion and 1 trillion T cells circulating in the tissues and bloodstream, any one of which can recognize only one target. The T cell receptors are so specific that they can kill an infected cell while leaving behind a benign neighboring cell. When a T cell recognizes an antigen, it makes more copies of itself while unleashing cytokines that marshal all the forces of the immune system to kill that invader. When someone gets mononucleosis, for example, up to 60 percent of all the T cells in the bloodstream are available to attack the virus.

Consider the immune system as the most powerful military machine ever assembled, with gunships, bombers, and tanks all equipped with savage weapons, all capable of killing enemies both familiar and unknown. Its awesome strength was seen during the worldwide Great Influenza of 1918, which, according to some scientists, came close to threatening the existence of civilization. This epidemic was not wiped out by doctors or researchers. It was defeated, at least in part, by the immune system. Once the influenza passed through a town, its residents developed immunity — their bodies roared back with their own defenses — and victims were not likely to be reinfected. (In time, the virus also mutated to a less lethal strain.)

But the Great Influenza also laid bare a dark reality of the immune system. The war analogy is apt: in any war, the military can accidentally kill its own with friendly fire, and the risk is higher when overwhelming force is used. This occurred, immunologically, in 1918. While immunity helped to stop the Great Influenza, it was also responsible for many of its deaths. As John Barry wrote, "The virus was often so efficient at invading the lungs that the immune system had to mount a massive response to it. What was killing young adults a few days after the first symptom was not the virus. The killer was the massive immune response itself."

Immunological friendly fire lies at the heart of type 1 diabetes. The adaptive immune system randomly generates billions of different T cell receptors, far more than the number of potential known enemies or antigens. It has a redundant capacity to destroy, which means if an asteroid were to hit the earth carrying an unknown bug — or if another pandemic influenza were to hit — it's likely that the body has a T cell receptor that will recognize and kill it. It's a beautiful, prodigious war machine. However, this redundant destructive capacity comes at a price. The system generates so many surplus T cells that some of them escape their intended purpose and attack healthy cells. This forms the basis of all autoimmunity. It may be tragic for the individual victims, but the immune system is so advantageous, indeed necessary, for the species' survival that sacrificing 1 or 2 percent of the population is acceptable to ensure that everyone else lives. Evolution doesn't care whether your child or mine lives or dies. It only cares that the population endures. This is no consolation to anyone with autoimmunity, no comfort to a parent whose diabetic child seems all but helpless against the vagaries of a heartless condition. But those very youngsters deserve recognition for their special sacrifice. It is no exaggeration to say that diabetes is the price we pay for the survival of all humanity.

What I wanted, of course, was impossible. In wanting a cure, I dabbled in the hubristic waters of modern science. I wanted to protect Garrett, to deny this disease its natural course, to cheat destiny. But I also wanted to defy 300 million years of evolutionary science, and I could no more do that than I could stop an ocean tide with the palm of my hand. René Dubos, a microbiologist who wrote *Mirage of Health* (1959), said that physicians and patients alike have historically overestimated the ability of human intervention to alter lives for the better: "The illusion that perfect health and happiness are within man's possibilities has flourished in many different forms throughout history. [But] complete and lasting freedom from disease is but a dream remembered from imaginings of a Garden of Eden."

Diabetes has been called the devil, but I suspect we have miscast the demon, even when it attacks our children. Though we want to protect them, we are better served by acknowledging our own limitations and recognizing the brutal truth of existence. The greatest disease of all is not diabetes or cancer or AIDS, or any other infectious enemy or immunological infidel. The greatest disease is time. "The problem is that we are flesh and blood, and our systems are bound to fail," Michael Bliss said. "Disease, aging, and deterioration win, so why should we be surprised that we can't easily conquer one form of deterioration? In the end, you can't win."

CHAPTER 14

Survivor Tales

WHY DO SOME DIABETICS flourish? There is no one answer, of course, but some themes do emerge. While successful patients understand that the burden is ultimately on them, they rarely fly solo. Instead, they find strength from many sources, be it their faith in God, their love of a spouse, or their support from family and friends. Many are motivated by the simple belief that health is better than sickness because life is better than death, but they attach their survival to a larger purpose, a greater good. They are also resourceful, finding creative ways to care for themselves in difficult, even extreme circumstances. Some remain angry, others frustrated; but most have kept their sense of humor. Too often only the failures are recounted, but the successes are proof of what is possible, not what is feared.

Günter Spiro's childhood shattered in early November 1938, when gangs of young Nazis roamed through his Jewish neighborhood in Berlin, breaking windows, burning synagogues, and looting homes and businesses. After *Kristallnacht,* Jews could no longer attend German schools, so ten-year-old Günter was sent to private school in Switzerland. His father, a decorated soldier from World War I, was stripped of all his assets and sent to the Dachau prison camp; he gained his freedom and fled the country with his wife. They reunited with Günter in London in 1940. From there, the family took one of the last passenger vessels to cross the Atlantic during World War II. Many of their friends and relatives perished, including Günter's grandmother, who died at Auschwitz.

The Spiros settled in Kew Gardens, New York, and tried to assimilate into their new country. Günter added "Robert" to his name — he would be known as Robert G. Spiro — and entered Columbia College. But his father could not escape his own nightmares. In 1948, Robert went home to their apartment and found his father dead by suicide. The windows were sealed. He had gassed himself. Robert's neighbors had to pull the young man out of the building.

Spiro finished college and entered SUNY Upstate College of Medicine at Syracuse. He was already interested in diabetes — his grandfather had died from complications, a massive stroke, when Spiro was three years old. Spiro wrote an honor's thesis on the disease, drawing heavily on Elliott Joslin's belief in good control. Then, in his last year of medical school, he began losing weight; he was hungry and thirsty and was always going to the bathroom. He conveyed his self-diagnosis to the chief of the Department of Medicine, who told him that medical students often believe they have the diseases they are studying.

In this case, however, the diagnosis was confirmed. Given Spiro's life to date, he could have gone into a nihilistic rage, but he embraced diabetes as something controllable, a disorder for which complications had a rational (if not yet scientifically proven) explanation. Achieving near normal glycemia was unrealistic in that era, but through discipline, innovation — and the extraordinary help of his wife, Mary Jane — he thought he could get close. He began by taking several daily injections, an approach that had fallen out of favor with the introduction of long-lasting insulins. They supposedly provided "coverage" throughout the day. Spiro, however, correctly recognized that numerous shots of short-acting insulin more accurately replicated the response of a healthy pancreas. He also began weighing all his food — more accurately, Mary Jane began weighing it to ensure proper portions. Everything was measured, even pieces of fruit. "There is no such thing," Spiro said years later, "as a standard banana."

Mary Jane was a chemist, and they set up a small laboratory at home to test morning blood sugars — a complex chemical process that required a homemade microcentrifuge, 200 microliters of blood (or about fifty to a hundred times more than the drop needed for modern test strips), thick lancets ("almost like swords," according to

Spiro), and a test tube that was immersed in boiling water for twenty minutes. A single test took a half hour; Spiro obsessively recorded every number. While this regimen caused frequent hypoglycemia — he began sleeping with a glass of orange juice on his nightstand, a practice he has maintained ever since — he believed that low blood sugars were a small price to pay for tight control.

Spiro joined the Joslin Clinic in 1961 and became a pioneer in establishing the relationship between diabetes and kidney failure. Mary Jane joined the lab five years later, and they collaborated for more than thirty-five years, examining how high blood sugar could impair the kidney's membrane. Ultimately, they succeeded in establishing a biochemical basis for the very care that Robert was providing himself. The professional payoff came in several prestigious awards, but the personal benefits paid a far greater dividend. After fifty-two years of living with diabetes, Spiro has no complications, enabling him to fully recover from surgery for a congenital heart defect. At seventy-seven, he is slight but physically fit, with a white goatee, thick glasses, and a gentle disposition. Now he uses an insulin pump and five-second glucose meters, but he still records his blood sugar levels. And Mary Jane still weighs his food.

Sitting in his cluttered office, Spiro can afford to be whimsical about his disease. "If you cured me tomorrow, I wouldn't know what to do with myself," he says. That may be true, but he also knows some harder truths. "I think it takes courage to take care of diabetes," he says. "It's much easier to let go."

Many do let go, but Spiro didn't, and he acknowledges that his early years shaped his outlook: when your family, your friends, and your country have been devastated by one of the worst calamities in history — when your entire world seems out of control — controlling diabetes doesn't seem so daunting. For most patients the disease is a burden, but Spiro transformed it into an opportunity, even a gift. Diabetes was beyond the reach of malevolent forces; it was something that he alone could maintain. A fatal condition, with no viable therapies, would have been traumatic, Spiro says, "but with diabetes, it's fabulous, isn't it? You can do something about it."

Cathy Fisher was a fitness freak in high school, a long-distance runner who reveled in her own athleticism. She would stretch standing in the supermarket line, raising her leg high onto the conveyor belt and slowly bending over until she could feel the blood pulsate up her calf, her thigh, and into her gut. She avoided soft drinks because she had read that carbonated beverages "displaced the oxygen" in one's blood. Her toes, discolored with purple blood blisters and crowned with chalky white calluses, were trophies. She loved pain — her feet pounding against the concrete sidewalk, her sweat streaming down her face, her "accordion lungs" inflating with air. She craved the endorphins that set off like fireworks at the end of a run. On St. Andrew's cross-country team in Boca Raton, Florida, she would complete races exhausted, sometimes collapsing as she crossed the finish line, but often in first place. Blue ribbons lined the walls of her room, and trophies towered like forts on her mahogany bookcase. She ran to win, but she also ran for health and strength and for a self-image of perfection.

When she entered Duke University in the fall of 2003, she rejoiced at suddenly being on her own. She lived by herself, met new people, did new things. "I was given the freedom to read Marx and Poe during the week," she later wrote in a college essay, "and dance on tables on the weekends."

But things began to go awry. She attributed her blurred vision to the strain of reading, but she could not explain her cravings for Coke after swearing off it in high school. She chugged entire two-liter bottles at a time, the last drops trickling down the plastic sides, then she'd head for the vending machines for almond M&M's. Inexplicably, she lost weight, and she felt the hard stares of classmates, who assumed she was anorexic. Her strength continued to ebb until one day, in the shower, she could not lift her arms. With the water pounding her slumped ninety-pound body, she hoped her hair would magically lift the green shampoo bottle and wash itself.

At the student health center, the nurse initially told her she had anemia. On the second trip, the diagnosis was strep throat. A third visit revealed that she was suffering from depression as well as an eating disorder. Each time, she was given cherry cough drops and ordered to gargle with saltwater at night. Finally, on a cold Saturday in November, she returned to the center and was redirected, undiagnosed, to the Emergency Room at Duke University Hospital. She could no longer walk and could barely speak. Blood tests revealed the diagnosis. She was soon on a table attached to three IV tubes, which fell across her arms like spiders — "daddy longlegs camping out on my skin," she wrote. Two IVs pumped saline into her dehydrated body; another, insulin. She cried, of course, but they were "the kind of tears that flow silently . . . unhindered, effortlessly, noiselessly, unstoppably." She had always tolerated the runner's pain, had even become addicted to it, but this was different. "This pain wailed and kicked and convulsed like a wild animal," she wrote. "This pain was real, not the kind of a leg cramp or a strained muscle. It was the pain of brokenness."

She went through all the phases of a new diabetic — the roiling emotional world far removed from the cheerful medical commercials and glossy trade magazines touting watercress salad recipes. At first she felt guilt, the same emotion she'd experienced after losing a race to a slower runner or even winning but not reaching her best time. She wished that she had "caught" the disease from someone else or had inherited it from some ancestor with dysfunctional genes. That she couldn't blame anyone else intensified her own shame. Then came denial; she insisted that diabetes was not a disease because disease implies sickness and she was not sick and she was not incapacitated and she was going to live with diabetes and she was not going to die, and anyone who called it a disease would be cursed out. Convincing herself that what she had was merely a "condition" was her way of staying strong: "Self-delusion can be very appealing when your only other option is self-hatred."

Then came anger; she resented the doctors and nurses who tried to reassure her with white lies — "this just means a couple of lifestyle changes." Then came despair. "My body no longer works the way it should. It is no longer perfect. Only perfect things have true value. I am no longer perfect, I can no longer have value." She wondered if boys would ever find her attractive and feared she'd pass diabetes on to her children.

Finally came isolation: "The loneliness was a taste of what my life would be from that moment on . . . I like to study in groups . . . live for my friends and family. But part of me will always be alone. No one checks my blood sugar for me, no one knows how to calculate my insulin requirements, no one makes sure I eat right. This part of my life will forever exist on its own."

She could not reconcile her prediabetic state — a resting heartbeat of 57, a 5-kilometer time of under 19 minutes — with the intractability of her current condition. "Sometimes it can be hard to believe in the forever of something as happy as a relationship," she wrote. "It's impossible to believe in the forever of something you despise."

She lay in the hospital and wished her present life away. In her mind, she was living in her dorm, attending classes, running up hills, and dancing on tables, beautiful and strong and healthy again.

Released from the hospital, Fisher returned home to South Florida for an extended Thanksgiving break. Her mother took her to a diabetic support group, against her will, which only made matters worse. Some members had had diabetes for a few years, others for three decades, but they all seemed to share the same hatreds — hatred for high blood sugars and for lows, hatred for doctors and employers and insurers, hatred for parents and partners and friends, hatred for the entire world. Fisher was dismayed that even a thirteen-year-old girl hated everyone and everything. She had thought that her entire life was about to open up, but now she felt as though she had walked into a long tunnel with no light. After a half hour, she walked out, crying.

Several days later, a friend introduced Fisher to her older brother, who also had type 1. Dan was in his twenties, a college graduate, and a professional who traveled widely for business and pleasure; he had backpacked across Europe and had even run a marathon. But what impressed Fisher the most was his attitude. While she was feeling hopeless and handicapped, Dan exuded a passion for life and was determined not to be limited in any way. Oddly, it appeared that he had become more alive because of his condition, as if a robust lifestyle was an act of defiance. He was, in short, living the very life that Fisher had once envisioned for herself — athletic and striving, lighthearted and optimistic. Diabetes had not killed his spirit.

The moment was a revelation, but it didn't so much send Fisher on a new path as it returned her to her original course. She was still the long-distance runner, endowed with the very traits — discipline, resilience, and willpower — that she would now need for her health. She found a superb doctor, Dan Mintz, at the Diabetes Research Institute in Miami, took a diabetes management course, and went on the insulin pump. She identified what had most depressed her about diabetes — the feeling of limits. So she committed herself to doing things as if she had no limits: she ran more than ever to prepare for her first half-marathon, she studied for a summer in Paris after her sophomore year, and she plans to backpack across South America when she graduates. At Duke, she became less anxious about grades and spent more time with friends. Boys still found her attractive. She still danced on tables. What did change was her ideal of physical perfection. She is now savvier, wiser, and more realistic. "Perfect exists," she wrote, "but maybe only in its flaws."

Eugene A. Bennett follows the daily regimen of any tightly controlled diabetic. He tests his blood sugar four times, takes five injections, and carefully records his results. He exercises regularly and minimizes his carbohydrates in favor of protein and vegetables. He reads extensively on diabetes as well as health and nutrition. He warns his daughters about the disease, encouraging them to follow a healthy lifestyle. He fears complications.

It's what you would expect from a man who spent years as a special agent for the FBI. But it's not what you'd expect from someone living in the Buckingham Correctional Center in rural Virginia, someone convicted of one of the most bizarre crimes in the history of the state. In June 1996, Bennett was arrested and charged with attempted murder and abduction in what prosecutors said was a plot to kill his estranged wife in revenge for her lesbian affair with the mystery writer Patricia Cornwell. Bennett took a pastor hostage at gunpoint to lure his wife to a Virginia church. His wife, also a former FBI agent, showed up with her gun and fired a shot at Bennett, who then fled and gave himself up a few hours later. His lawyers asserted that he was not guilty by reason of insanity — expert witnesses said that Bennett had developed an evil alter ego that sometimes controlled him — but he was convicted on nine felony counts and sentenced to twenty-three years. Sensational headlines followed.

Addressing the court at the time of his sentencing, Bennett said his undercover work damaged him psychologically. "Leave it alone," he said. "I was one of the best. Now I'm a walking case history for the downside."

Two years later, he began losing weight and was feeling weak. Prison nurses and doctors said he had the flu — which persisted for nine months. Bennett continued to deteriorate until one day, as he lay immobile on his bunk, his cellmate thought he was dead. The guards had to carry him out, and an ambulance rushed him to a hospital. His blood sugar was 1,140, his throat 95 percent closed from a yeast infection. He woke up the following day to be told he had diabetes — another sentence, but this one for life.

Most diabetics who aggressively manage their condition have a basic survivor's instinct, but if survival itself is a joyless, suffocating grind, the endless sacrifices are more difficult to rationalize. Why prolong the misery? The question played out for Bennett, whose prison has about a dozen insulin-dependent diabetics. He noticed that many didn't take their daily shots. Of the prison's entire diabetic population, including another forty who took pills, many were overweight, ate poorly, and didn't exercise. They had surrendered to the disease and paid the price: some of the older diabetics had toes, feet, or legs amputated.

Bennett was in his early forties at his conviction. His career over, his reputation ruined, his assets gone, he could have followed that same path, allowing diabetes to ravage his body. Instead, he refused to succumb, raising "self-care" to remarkable levels. He didn't understand the disease, so he foundered for the first ten months, whipsawed by roller-coaster blood sugars. Receiving little guidance from the prison, he sent letters to diabetes organizations, doctors, and hospitals in search of information. When he began receiving books, pamphlets, and magazines, the assistant warden called him in and asked why he was receiving so much medical literature. "It's pretty obvious," Bennett said. "I can't get it from the medical department." When he ran out of money to renew his subscription to *Diabetes Forecast,* he asked the ADA for a free subscription. Two weeks later, it fulfilled his request and sent him twelve books as well.

Dissatisfied with the prison doctor, Bennett was able to use a Tele-Med system to consult online with a specialist at the University of Virginia Medical Center in Charlottesville. That service ended after several years, but by then he had learned the basics of tight control through intensive insulin therapy, diet, and exercise. Assisted by his lawyers, he received the best diabetic care in the prison. Most of the inmates took one or two shots a day using older, less effective insulins. Bennett injected himself five times a day with the newer insulin analogs, Humalog and Lantus. Bennett was also the only prisoner who determined his own doses.

Each inmate has his own glucose meter, donated by a medical device company, but the prison authorities don't allow spring-loaded lancets for finger pricks. Prisoners must therefore stab their fingers to draw a drop of blood — a painful exercise that discourages most of them. Bennett, however, tests four times a day. He is also fanatic about exercise, working out on the "weight pile," running three miles three times a week, and jogging up to ten miles on Sundays.

But maintaining sound control is complicated by many factors, starting with the food. Most meals are glycemic nightmares, consisting of inexpensive carbohydrates and fats, with relatively little protein. Overcooked vegetables are drained of nutrients; thick brown gravy often covers the entire plate. The commissary offers a grim smorgasbord of cheese puffs, Snickers bars, and lasagna pouches. But Bennett finds ways to circumvent the system. He earns 45 cents an hour tutoring other prisoners, and he uses it to buy junk food from the commissary while hoarding his own desserts (frosted cake, chocolate chip cookies, pudding). He then trades his sweets to other inmates for their chicken, peas, carrots, or bananas. When healthy foods are served, Bennett will use Ziploc bags to smuggle extras out and store in his cell — a risky gambit, as guards occasionally pat down inmates leaving the mess hall. He also buys food (rice, oatmeal, onions, powdered milk) from kitchen workers, even though the trades violate prison rules and could subject him to punishment.

Bennett runs a kind of credit store from his cell, giving a "good credit" prisoner two cans of Coke in exchange for three cans, due on payday. As long as the default rate does not exceed 20 percent, Bennett earns a profit, for he can sell his extra cans to other prisoners. In 2004, a new program allowed inmates to order packaged food, which — while more expensive — permits Bennett to buy tunafish, salmon, and dehydrated refried beans. His greatest stress occurs during "lockdown," when, after a fight or some other threat, inmates can't leave their cells (Bennett's is six feet by nine), nor are they allowed even to exercise in their quarters. Lockdowns last for up to two weeks. What's more, his food and his insulin are not distributed in sync, wrecking any balance to his schedule. For the most part, he's avoided severe hypoglycemia, though one night he barely made it to his locker, eighteen inches away, to drink his Coke. Now he sleeps with Coke and M&M's in his bunk.

Bennett tried to round up a diabetic support group but had no takers. He can't understand why so many prisoners skip their shots, "fail to even walk outside at recreation and eat themselves silly," or why the "superfat boys" angrily claim the glucose meters are broken when their blood sugars are sky high. ("The nurses just roll their eyes and send them on their way with a huge dose of insulin.")

But a better question is why Bennett himself goes to such lengths to maintain his health. Like any diabetic, he has moments of exhaustion and despair. "Some days," he says, "I feel like just saying 'fuck it all' and eat what I want, sleep all day, not go to work, not get a finger stick and no more damned needles." But he knows the price of surrender: "They can kiss my eyes, kidneys, liver, heart, feet, legs, etc., good-bye in a few years. So I get over it and get on with it."

Self-preservation, however, is not his only motive. Anger is another. He said that his presentence report recommended 6½–7½ years — which he described as "pretty fair. Punishment enough to get my attention, [to] get my head unscrambled . . . and still have a chance to be a good citizen." But the court initially sentenced him to sixty-one years, only to be reduced on appeal to twenty-three. "So I'm gonna be sixty-two, sixty-three before I get out of here with no family left, few friends, no assets, and fewer options. So my only way to survive and not give in or give up is to stay healthy and stay in the best possible shape I can till then . . . I'll not die in prison." Survival is his only revenge, his way to defy "all my former fair weather friends and family members who bailed on me when I needed them most. I know this is the negative side of motivation, but it works for me."

The most remarkable story in the history of diabetes belongs to Eva Saxl. It is a sweeping tale that combines wartime miracles in a refugee camp, the scientific improvisation of a beloved husband, and the abiding faith of an unassuming linguist who defied all odds.

Born in 1921 in Prague, Czechoslovakia, Eva Saxl knew adversity at a young age. Her vision was poor, so her parents hired special tutors to train her in different languages. She eventually spoke eight, honing them at finishing schools in England and Switzerland. At thirteen, Eva told her mother she intended to marry a distant relative named Victor Saxl, and she made good on her promise six years later. By then — 1940 — the Nazis had invaded Czechoslovakia; the newlyweds fled to Genoa, Italy, where they boarded the *Conde Verde* with about three hundred other Jewish refugees. The vessel was the last ship to pass through the Suez Canal during the war. The worldwide depression had caused many countries, including the United States, to suspend immigration, but refugees were still accepted in what became known as "the port of last resort" — Shanghai, China. Eva and Victor settled in the Jewish ghetto, an area of less than one square mile for about 18,000 Jewish refugees.

It was hardly a safe haven, for the city had been controlled by the Japanese since 1937. Victor, a textile engineer, managed a large woolen mill factory, and Eva found a job teaching languages. One day Japanese soldiers descended on the school, lined up the teachers, and began firing. According to Eva, she fainted before she was hit and was left for dead. She woke up and was embraced by the students.

Next a very different crisis arose. She began losing weight and craved water, so she visited the school's doctor, who gave his diagnosis: *"Sz tan cze bin."* Eva understood: "It is the sugar water disease." Insulin, from Eli Lilly, was readily available, and Eva made the necessary adjustments. Victor gave her an injection each morning but insisted that she do it at night to keep in practice. "His love helped me avoid the psychological problems of diabetes," she later recalled. But his love was not enough after December 7, 1941, when Japan attacked Pearl Harbor, causing its army to tighten its grip on Shanghai. Pharmacies were closed, and the International Red Cross could not bring in medical supplies, forcing the diabetics in the Jewish quarter — more than four hundred — to scurry through town buying up all the insulin available. Some patients, including Eva, tried to buy it on the black market, using one-ounce gold bars for payment. But the insulin's safety was not assured: one of Eva's friends died after his first injection. The liquid had been poisoned.

To reduce their need for insulin, the diabetics went on modified starvation diets. They also tried Chinese herbs as well as German Synthalin — tablets that supposedly lowered your blood sugar but were neither effective nor particularly safe. Eva called them "revolting little silver pills." The weeks passed, and Eva monitored her dwindling cache of insulin; it was stored in boxes cooled by expensive towel-wrapped ice blocks. Each bottle represented so many weeks of life. Some diabetics had already run out and were hospitalized in coma wards. Others died. To minimize Eva's panic, Victor secretly refilled the empty insulin bottles with water and milk powder to make her believe she had more "life" left.

Lack of insulin was only one problem. The diabetics used glass syringes that, if broken, could not be replaced. Eva owned several that she would sterilize daily over coal-dust kitchen fires. They were chipped, scarred, and discolored, but they never broke. Needles, also scarce, were dulled with continued use, making injections more painful.

From the moment Eva was diagnosed, Victor had pledged to take care of her, but by early 1943 he knew that desperate measures were needed. With no hope of importing insulin, only one possibility remained: they had to make their own. He met with a group of internists and pleaded for help. They refused — they were afraid of creating a lethal batch that could ruin their reputations — but they gave him their medical books, which included some reference to the making of insulin. The volumes themselves came in six different languages, which presented no problem to the married linguists. (Victor was also fluent in several tongues.) The most useful book was *Beckman's Internal Medicine,* which included drawings of Banting and Best's dissection of dogs.

Victor next found a small laboratory in the Municipal Building, where a Chinese chemist tested food and drink samples from street vendors. Mr. Wong agreed to let Victor use his primitive lab and to help him as well. So a textile engineer and a food chemist set out to make insulin. Needing fresh animal organs, Eva and their cook took a rickshaw to the Seymour Road Slaughterhouse each morning at five o'clock to claim the pancreas of any large animal. Those of water buffalo were the biggest and best, blending nicely with pig pancreata. Eva stuffed the bloody masses into a wide-mouth thermos flask and hauled them back to the lab. There, they were ground up using Eva's mother's old kitchen meat grinder — Eva herself sometimes did the grinding — then were quickly heated and cooled to kill the organs' enzymes. Also needed were alcohol and blocks of ice, which were provided by diabetics who were not confined to the Jewish ghetto.

Victor's experiments progressed slowly. He tested his concoction on six rabbits, each starved for twenty-four hours and then divided into two groups. One group was injected with his mix; the other, with "real insulin" donated by Eva and two other patients. Victor did not have the equipment to test the rabbits' urine or blood, so he waited to see if the animals receiving his homemade variety experienced the same hypoglycemic shock as the other rabbits. The time difference allowed him to measure the potency of his solution. Eva herself couldn't watch the experiments; she became too upset when one of the rabbits died.

The experiments were a poor substitute for human testing, but when Eva had only five days' worth of insulin left, Victor had no choice but to try his concoction, once again using a ruse to calm his wife. He rushed home with a Czechoslovakian doctor and told Eva the good news. "I've got some new Japanese insulin for you," he said. Eva saw through the caper. The Japanese had not given them insulin in two years, and the solution in Victor's hand was brown. When she asked about the color, the doctor explained that the Japanese were now making fish insulin, which could hurt more.

Victor's shaking hands filled a syringe. "Shall we try it?" he asked.

"Yes," she said. "Let's try the fishy stuff now."

Victor injected the murky fluid. With tears in his eyes, he told his wife he loved her and ran out of the room. The doctor prevented Eva from following, saying, "Let Victor pray. Let us pray too." He kept an eye on his watch, and Victor returned with the cook, who carried in a tray of tea. The doctor said he had some saccharin, dropping it into the cup. But it didn't sweeten the tea, and Eva knew it was a sedative. The time passed without any side effects, and soon Eva began to believe the insulin was working. She told Victor she felt better. She recalled, "We were crazy with happiness."

They rushed over to the hospital to give the insulin to two dia-

betics lying in coma. The patients recovered; one named his son after Victor. The other diabetics in the ghetto signed up for a supply, bringing money for black market alcohol and other expenses.
Some arrived with blocks of ice. One brought his needle sharpener
to the lab — it looked like a pencil sharpener, with two flat, round
grindstones on a stand and a handle to move the stones. People
would sharpen their needles when they picked up their insulin. Victor also raised money by making woolen stockings. Mr. Wong continued to test food and beverages, next to the small insulin factory,
and never asked for payment — though Victor did give him some
socks.

The insulin itself was made in small batches. About 50 percent
had to be discarded, and patients were told not to use more than sixteen units a day. But shortages of alcohol and ice were chronic, the
electricity would often fail, the money could not keep pace with rag-
ing inflation, and the Allied bombing raids terrified everyone. The
bottles proved to be another problem. When Victor's insulin was put
in empty imported bottles, the product was fine, but when it was put
in bottles from a nearby factory, it spoiled quickly. No one understood why. The Saxls asked the diabetics to bring in all their empty
imported bottles. They were amazed that few had been thrown out,
leading Eva to conclude that her fellow diabetics were so desperate,
they even clung to their empty vials.

Not a single diabetic died using Victor's insulin over the next two
years. When Shanghai was liberated, the Saxls asked a U.S. Marine
medical officer for some "white insulin," the clear, purified variety Eva had previously used. When the officer asked what type and
strength, they burst into tears.

After the war, President Truman signed a bill allowing people
with special abilities to enter the United States, and the Saxls qualified. When they arrived in San Francisco several years later, Eva
walked into every drugstore she could find just to ask if they had insulin. The Saxls were told they should travel before settling down, so
they took a train across the continent, stopping to see Truman in
Independence, Missouri, just to convey their gratitude. "We wanted
to thank Mr. Truman for letting us into America," Eva later explained. "He said no one had ever done that." The president invited

them in for tea, and they listened to his daughter, Margaret, play the piano.

The Saxls then traveled to New York. Still a young couple — Victor was in his thirties; Eva, her twenties — their lives took another amazing turn. When Eva saw a doctor for a routine checkup, she mentioned how Victor had saved her in Shanghai. The doctor promptly called the ADA, which recognized the potential publicity coup: a spokesperson for diabetes who could testify to the miracle of insulin.

With their irresistible story, Eva and Victor became celebrities. For many years they traveled across the country, giving speeches and appearing on numerous radio and television shows, including Edward R. Murrow's. Eva also contributed an essay for Murrow's book *This I Believe*. Hal Roach Studios, better known for producing Laurel & Hardy films, made a television documentary about the brown insulin and the water buffalo. The Saxls were invited to the White House and met President Eisenhower in the Oval Office. Charles Best invited them to speak in Toronto, and Victor delighted in telling the assembled scientists how he and a Chinese food chemist reinvented insulin. The following night, the Saxls spoke at a hospital while hundreds more waited outside to hear the same speech. The couple met Elliott Joslin, who taught Eva about diabetes care and lent her medical books. Joslin also asked Eva to tell her story to various groups, particularly diabetic children.

Her celebrity was unusual at a time when most diabetics, rightfully concerned about discrimination, meticulously hid their condition. While she proved her resilience in Shanghai, her willingness to discuss her experience was an even greater measure of her character, as she must have inspired other patients bereft of role models. But the publicity also probably fueled misperceptions about the disease, for it did nothing to convey the realities of diabetes and its complications. Rather, it further embellished insulin's magical powers. Even impure insulin, created by scientific neophytes in a crude lab, saved lives.

In 1968, Victor had been working for a textile firm in New York when he accepted a position with the United Nations to provide technical support in developing countries. Accompanied by Eva, he went to South America on a three-week assignment. While visiting Eva's brother in Santiago, Chile, he suffered a heart attack and died; he was fifty-eight. To Eva, the death was not just agonizing but inexplicable. Victor was a regular tennis player and swimmer who had always shunned tobacco and alcohol. He had spent twenty-seven years caring for his "sick" wife, calling her every day, giving her injections, and ensuring her health. Now Eva was a widow at forty-seven, overwhelmed by loss and fearful that she could not go on. Drawing on prayer, she concluded that she should live the kind of life that Victor himself would have aspired to — a life of helping others. She needed a mission and didn't have to look far to find one. Her "silent partner" had long been diabetes, and she dedicated herself to that cause.

Staying in Santiago with her brother, she taught at the Diabetes Association of Chile, the Information Center for Chilean Diabetics, and at Red Cross centers. Her travels now took her across the Americas and, indeed, the world, her most dramatic visit being to Düsseldorf, where she spoke at the German Diabetes Association Congress. She also visited her native Prague, where she was made "first member of honor" of the Czechoslovakia Parents and Friends of Diabetic Children. She walked through her old neighborhood but said, "Without loved ones, it's not the same."

Her story appealed to new generations. In 1991, she appeared on the cover of *Diabetes Forecast,* was featured on CNN, and received the Charles H. Best Medal "For Distinguished Service in the Cause of Diabetes." The awards ceremony in New York was attended by 1,500 people.

In the summer of 2004, Michael Bliss told me the outlines of Eva Saxl's story. He had met her some years before and still had her telephone number in Santiago, but he assumed she had died by now. I asked for the number anyway, thinking that whoever lived there might know where I could contact her friends or relatives. A month later, I called. A woman with a melodic voice answered after the first ring.

I introduced myself and said I was writing a book about diabetes. I was a bit nervous, not knowing how to inquire about a dead person, so finally I just said, "I'm trying to find Eva Saxl."

"This is Eva," she said. "I'm so glad you called."

I was stunned but at least had the good sense not to say, "I assumed you were dead." She told me she lived in an apartment with her maid and caregiver, Clarita. She had had some health issues in recent years, including surgery to remove a cancerous colon. "My tummy looks like a map of Westchester," she said. She also had had cataracts removed from an eye, but the surgeon said that in all his years of operating on diabetics, he had never seen a retina "so untouched by any complication."

Eva said that diabetic care had become much easier once glucose meters were available, and some years ago she had adopted a strict low-carbohydrate diet. Her excellent control had enabled her to withstand the surgeries and other health problems. Though legally blind from her prediabetic condition, she said she could still see "well enough."

Talking to Eva was like meeting a long-lost friend: consoling and encouraging, she was enthusiastic about my work, inquisitive about my health, and bullish about the advances in medical care. The keys to Eva's own survival emerged. She never allowed any physical ailment to eclipse her irrepressible optimism, while her abiding love for her husband, even years after his death, remained strong. Victor's sacrifices and commitment sustained her during her darkest hours, and now they were a beacon that guided her through her twilight years. Finally, Eva's religious faith was a constant source of comfort and renewal, a belief that God would not betray her.

I didn't want to tax Eva further on this call, so I asked if we could speak again. She said of course, and we agreed to talk in a few months. Before saying good-bye, I asked if she had made it to temple during the recent Jewish High Holidays.

"Oh, no," she said. "I can't get out like that anymore, but that's okay. I have a direct pipeline to God."

Several weeks later, Eva Saxl died. She was eighty-three.

Epilogue

I return to Orlando one year after the ADA convention. It seems like a lifetime. I've returned not as a writer but as a participant, and I've brought the entire family. We're attending the Annual Friends for Life Conference and Expo; it's sponsored by Jeff Hitchcock's organization, Children with Diabetes. Sweltering heat deters most families from visiting central Florida in July, but this conference is worth it. It began six years ago, when the organization e-mailed parents to see if anyone would like to share a vacation in Orlando. About 550 people showed up. This one has drawn 2,200, including more than 1,000 children, from forty-two states and ten countries. I'm not sure if the epidemic is worse than we know or the need for information is greater than we realize — I just know that Jeff has built a powerful community of kindred spirits.

The four-day event occurs at Disney's sprawling Coronado Springs Resort, whose jogging trail wends around a fifteen-acre lake. The conference's motto is "Kids Being Kids." There are fun runs, scavenger hunts, face painting, volleyball, and swimming. Mickey Mouse, the Beauty and the Beast, and lesser princes and princesses drift in and out of events. Teenagers crowd into the Fiesta Ballroom for an evening dance; a glucose meter is required for admission.

Inspiration abounds, as hale diabetics describe how their condition does not prevent them from realizing their dreams. The mountain climber Will Cross, last seen a year ago shaking hands in the Novo Nordisk estate, talks about his ascent of Mount Everest. He climbed 27,500 feet above sea level, just 2,500 feet short of the summit, but had to turn back because of exhaustion and insufficient oxygen. (He says he'll try again.) Jay Hewitt — a corporate lawyer and part-time fashion model — regales us with equally amazing feats: his competition in iron man races that call for 2.4 miles of swimming, 112 miles of biking, and 26.2 miles of running. Douglas Cairns, a former jet pilot for the British Royal Air Force, describes how he flew around the world in a light aircraft in 159 days, making sixty-three flights over 26,300 nautical miles. And the former Miss America, Nicole Johnson Baker, radiant as ever, shares her experiences as a pageant queen.

The speakers are all strapping, successful, and attractive, but I have no quibble with such an idealized tableau. While Garrett is too young to appreciate it, the older kids surely benefit by hearing that other diabetics have not been limited. "Dream big and go far" is a good message for anyone, and it's repeated throughout the conference. Before breakfast, as the throngs line up outside the banquet hall, someone leads a cheer: "Today I will so tomorrow I can!"

Of course, diabetics often get mixed messages, and the conference is no exception. On the last day, participants go to the Magic Kingdom and are given, courtesy of Disney, Guest Assistance Passes, intended for visitors with special needs — in our case, diabetes. The passes allow us to skip lines and go directly on any given ride. So if you have diabetes, you can climb Mount Everest, compete in a triathlon, and fly around the world, but you can't wait in line for Pirates of the Caribbean.

Sheryl and I attend several breakout sessions. One of the most popular is led by Richard Rubin, a soothing, silver-haired psychologist who has both a sister and a son with diabetes. The title of his speech resonates: "Overcoming Diabetes Overwhelmus." He describes the old days, when you would "test your urine to see what your blood sugar was." Everyone laughs, myself included, though I didn't realize how funny it was until now. He tries to reassure parents of newly diagnosed children: "We think our job is to control our kids' blood sugars, but our real job is to help our kids learn to control their own blood sugars." We need three things to cope, he says: "love, faith, and humor." He then tells a story. "When a twelve-year-old girl was diagnosed, she asked her mother what this meant. The mother said, 'We're going to have to be more active than we've ever been before. We're going to have to eat healthier than we've ever eaten before. And we're going to have to love each other more than we've ever loved each other before."

We attend a speech by Fran Kaufman, one of the patron saints of the field, who shares an anecdote from her book, *Diabesity.* She recounts that as the head of the Center for Diabetes, Endocrinology and Metabolism at Children's Hospital Los Angeles, she received a late-night call from the hospital that a nine-month-old boy named Cameron lay in a diabetic coma. Kaufman's husband was out of town, so she had to grab her two children, ages three and almost one, and rush to the hospital. In the Emergency Room, as Kaufman tried to revive Cameron, her younger child began to whimper. He then threw a juice bottle across the room and cried uncontrollably. Cameron's mother, visibly upset, finally approached her.

Kaufman reads from her book: "My child was crying hysterically, but I spoke as calmly as I could, trying to look as if I had the situation under control. Cameron's mother knew her son's life was in danger. She was terrified and exhausted and I wanted desperately to reassure her. But Jonah was inconsolable. I felt frustrated and embarrassed. 'I'll tell you what,' she said, 'I'll hold your child if you save mine.'"

But Kaufman can't quite finish the sentence. She chokes up, her eyes tearing, still deeply moved by the image of two mothers in need. She apologizes to the audience, but it's unnecessary. Most are crying with her. And she saved the boy.

The drug and medical device companies are here, sponsoring events, meals, and snacks, all of which keeps the price down for families. (The registration costs about $300 per family.) The Expo, with vendor booths and product displays, is not as gaudy as the ADA's carnival, but fun and games are still part of the mix. Once again, Novo Nordisk has the most elaborate setup, with a "rock wall" that kids can climb, while Accu-Check sponsors a flight simulator used by Michael Hunter, a conference speaker described as "the only known insulin-dependent aerobatics pilot in the world."

But a more serious side of the disease is also evident; not surprisingly, it comes from the children and their parents. Hitchcock's organization sponsors a Quilt for Life, comprising three-foot-square pieces submitted by different families, creating a vast tapestry of the diabetic experience. One piece includes a boy juggling a basketball, baseball, and tennis racquet (for food, exercise, and insulin), walking across a tightrope above water, with sharks circling below. Another has a photograph of a boy and is inscribed: "Diagnosed with diabetes on January 17, 2000, at age seven . . . He has had over 5,000 finger pokes to test his blood sugar." Another has a hospital record of a boy's A1c numbers, with a poem by his parent:

> A thousand tears
> A thousand prayers
> I'd give my life away
> To have my son be healthier
> For just a single day.

These sentiments are far removed from Garrett's experience. He is now four years old, and he's swimming, running through the spacious corridors, and flinging pennies into the hotel fountain.

"What do you wish for?" I ask him.

"That I never leave," he says.

I wish I could say that Garrett's first year with diabetes was a breeze, that he accepted his routine, and that he always cooperated. That would be a lie, though his first year was probably no worse than most and better than many. On the positive side, and most important, his health was good. His A1c's were 7.4 and 6.8, which is very good for a young child under any circumstance, but particularly for someone taking only two shots a day. We need not worry about his growth being stunted: in one three-month stretch he grew an inch, and he'd be the center on his basketball team except for Olivia, who's got him by a smidge. Growing that much complicates his insulin needs, since his body is constantly changing. But that's a good problem.

Garrett understands that diabetes is about controlling his blood sugar and that glucose readings give us the information we need to make adjustments.

"High or low?" he asks us after the meter beeps.

If we say low, he pumps his fist and says, "Yes!" He knows he'll get a bigger snack.

We've turned a central part of his care into a game. Researchers now know that even if average blood sugars are close to normal, glycemic variability — or sharp swings — can be damaging, and these excursions usually occur right after eating. Two possible ways to minimize the spikes are to significantly reduce your carbs or to inject your insulin about a half hour before a meal. Neither of these is possible for Garrett. He eats mostly carbs, is disdainful of virtually all protein and vegetables, and once he receives his shot, he wants to devour his food immediately. That leaves a third possibility: exercising after eating. You don't have to run a marathon; even a little movement — washing dishes, for example — can be helpful. What you must avoid is eating a big meal and then sitting on the couch for two hours. When Garrett loads up on carbs, I tell him we'll have to do wind sprints. Sometimes he'll volunteer, asking for another slice of pizza and promising (if mispronouncing), "I'll do wing sprints! I'll do wing sprints!" And off we go. If we're at a restaurant, we'll leave Sheryl and Amanda behind and do sprints in an alley, on the sidewalk of a strip mall, or along the main street of our town, sometimes even in the rain. If we're in the house, we sprint in our playroom — ten, twenty, thirty dashes back and forth.

"He's got quite a motor," his soccer coach tells me.

"Yes, he does," I say. "Yes, he does."

We've been fortunate. His preschool teachers, one of whom has diabetes, are attentive and loving. They touch his forehead when he naps, buy him sugar-free Popsicles in lieu of real ones, and promptly recognize behavior changes that signal trouble. On a few occasions, when Garrett was lethargic or simply stood alone against a wall, they knew something was wrong and gave him a snack.

But we weren't always as careful as we should have been. When his preschool was closed for several days, we enrolled Garrett in a day sports camp. I dropped him off and told his counselor that he had diabetes and that we had packed him his snacks. I asked what time the kids had their snacks, and the counselor said 10 A.M. and 2 P.M.

I figured Garrett was safe. I left him off at nine-fifteen and would be back at eleven-thirty to test him. What could go wrong in a few hours?

At 11 A.M., the telephone rang. It was a woman from the camp.

"Garrett is crying and we don't know what's wrong," she said. "He won't stop."

"He has diabetes!" I screamed. "Give him some juice now! He has to drink it!"

I grabbed his meter, an apple juice box, and, in case he was passed out, a glucagon kit. I realized that I should have specified on the telephone that Garrett needed regular juice, with sugar — not the Kool-Aid Jammers 10 juice that Sheryl had packed for him. Each pouch has only 10 calories and 2 grams of carbohydrate, which is good for maintaining blood sugar levels but not raising them. When I arrived at the camp, all my fears were confirmed. Garrett never ate his snack — they were playing basketball instead, and the counselors forgot. And they didn't give him regular juice but two pouches of Jammers.

"He just sucked them down," one of the counselors said.

The sugar in the pouches, however minimal, kept him conscious, and perhaps counterregulatory hormones kicked in as well. He was still shaken, still sobbing. But when I tested him, he was 82, and I gave him some cookies.

"He just started crying," the counselor said. "Kids cry all the time, but then he couldn't walk off the court. He kept falling down."

Yes, that's what happens, I thought. You fall, and you fall some more, and you keep falling.

I reminded the counselor that Garrett had diabetes and he needed to eat his snack.

"Oh, yeah," he said. "I didn't put two and two together."

I was tempted to yell at him, but it was my fault. I had spent time with Garrett's preschool teachers, explaining the disease and how he needs food to counteract his insulin. I had done none of that here. I took Garrett home; he was still upset, still frightened. And low. His blood sugar was back down to 53, but he could eat his lunch and was soon stable.

That's the risk of tight control. The margin for error is very narrow, and missing one snack can have horrifying consequences. We've told Garrett the symptoms of being low, but he still can't recognize them. Age isn't the only problem. If he ran chronically high blood sugars, he'd be more sensitive to lows, but his relatively good control renders him asymptomatic when he's in the sixties or even the fifties. This is another way the disease punishes compliant patients. By the time Garrett does feel it, he's in real danger — and defenseless.

I recall the story of the girl who slipped on the icy sidewalk. A man rushed to her and asked, "Can I help you?" "Yes," she replied. "Catch me before I fall." That's what all parents must do, except the diabetic child's fall is so much steeper. I had left Garrett on the precipice; he slipped and was heading down the cliff. We pulled him back, but we can't afford to let him slip again.

Amanda continued to impress us with her medical attentiveness, not just tracking Garrett's blood sugar numbers but learning about the disease by overhearing conversations between Sheryl and me. When Garrett told us that a parent of one of his teachers has diabetes, Amanda asked, "Type 1 or type 2?"

I looked at her, amazed. "How do you know the difference between type 1 and type 2?"

She shrugged. "I just do."

Then Garrett surprised me. "I have type 2!" he exclaimed.

"Why do you think you have type 2?" I asked.

He pondered the question. "Well, I have it in the morning and I have it in the evening."

The hardest part of Garrett's first year, by far, was the shots. We know other children who don't resist them, but Garrett did. Perhaps his age was the problem. Diagnosed at three, he was young enough to fear needles but old enough to fight back. Following advice from the experts, we tried to divert him or involve him more in the process, but everything we tried made things worse. We were told to give Garrett a ball or stuffed animal to squeeze. He has a few favorites, which we inevitably misplace, allowing him to demand that we find his "squeezie" before we give the shot, thereby delaying the inevitable. I would also remove the caps of four syringes, hold them like four pencils in my hand, and ask Garrett to choose one. He meditated over which was shortest and chose. Of course, each needle was the same, but it gave him a chance to stall.

The shots themselves still elicit occasional outbursts of anger, of hits and kicks and screams of "I hate you!" or "I don't want my diabetes anymore!" Sheryl is more patient than I. "He's just frustrated," she says. Regardless, the eruptions are short-lived. The shot is done, the hostility subsides, and he's quickly laughing or playing again. While the episodes are unpleasant, part of me hopes he maintains that edge, that defiance. People with diabetes should be angry — not at themselves or the disease but at the entire health and medical field, holding all the players accountable and demanding more than what they're getting. Complacency is no virtue.

The other major hurdle has been food. Carbs notwithstanding, his eating is actually reasonable, and he doesn't crave sweets or binge. The problem is that as a growing boy, his favorite sentence is "I'm hungry." We know this makes him normal. Growing boys eat. That's their job. When his friends come over, regardless of the time, they raid our snack cabinet. The kids aren't overweight or gluttonous. They're just growing boys.

One solution is the insulin pump, which allows patients to more readily cover snacks or meals at irregular hours and is becoming a therapy of choice for children. Sheryl runs a playgroup for diabetics under the age of seven and their families. Of the forty kids, about a third are on the pump, and we know it's just a matter of time for Garrett. Nonetheless, part of me resists, not just because I know the long infusion set needle will petrify him. I also have this unreasonable concern that the pump will entangle him, both physically and emotionally. The way he plays sports or just plays with friends, hurling his body into the maelstrom, does not seem compatible with a small machine attached to his body. And maybe, deep down, I don't want his lifeline exposed so that people see him as different or vulnerable. When I test his blood sugar at his preschool and another child asks what we're doing, I sometimes hesitate, mumbling something about how it's just something we do because it's cool. Even at such a young age, I want him to be accepted for who he is, not for what he has.

We may need to put him on a pump for another reason. Eli Lilly announced in 2005 that it was discontinuing the production of its Humulin Lente insulin, which was the long-lasting insulin that worked best for Garrett. The company also said it would stop making its Humulin UltraLente. Combined, the insulins were used by about 66,000 patients. According to Garrett's nurse at Joslin, they were mostly used by children, but Eli Lilly said its sales had declined due to other treatments. Before the year ended, we got extra prescriptions for the precious Lente and hoarded vials in our refrigerator; I thought of Eva and Victor Saxl storing bottles in Shanghai. If we run out before Garrett moves to the pump, we'll have to find an inferior alternative. Eli Lilly's decision was a useful reminder that children are the most underserved customer in this disease; and for all the efforts by drug and medical device companies to improve the lives of diabetics, the day the illness no longer makes money for them is the day they get out.

Ironically, products that appeal to diabetic children should do well, because their numbers are growing, and the success of Jeff Hitchcock's organization is only one indication. With Garrett several months away from kindergarten, Sheryl and I met with the Needham School District's nurse supervisor. Four years ago, she said, the district had two children with type 1 diabetes. Now there are fifteen! She had no explanation for the increase.

The type of insulin, ultimately, will not determine Garrett's destiny. It will be his attitude, and I will tell him what Joslin's Howard Wolpert tells his patients: Be realistic. The goal is not perfection but perseverance. When you have a good day, take the credit. When you have a bad day, blame the disease and vow to make better decisions tomorrow.

The good news from this year is that Garrett himself didn't change. Notwithstanding his resistance to injections, he didn't mope about the disease. It didn't make him sullen or spiteful. He remained an affectionate, energetic kid beloved by teachers and coaches, a playful brother, a loyal friend, and a mostly dutiful son with an independent streak that will serve him well. When he sees Sheryl or me doing laundry, vacuuming, or barbecuing, he often demands, "I want to do it!" He wants to do it by himself and gets angry if he doesn't do it perfectly. I hope that Garrett takes care of his health with the same determination and self-sufficiency. His life depends on it.

On our last full day in Orlando, we wear our lime-green "Friends for Life" T-shirts, slip on our "Insulin is not a cure" bracelets, and board a shuttle bus for the Magic Kingdom. We meet a five-year-old boy, Zachary, and his mother, who tells us they live on an island in the Seattle area. She knew diabetes was serious, she says, when her son was diagnosed, because the doctor told them to hold the ferry. "Normally, they don't even hold the ferry for pregnant women," she says.

Garrett and Zachary are talking, swinging on the metal poles, pushing and squeezing. The boy is wearing an identification necklace with a small baseball on which the diabetes insignia is printed. Zachary tells Garrett that he takes the necklace off at bedtime, and Garrett shows him his bracelet. He then looks up at me with his big eyes and asks, "Daddy, when can I take off my bracelet?"

At this point, a story about diabetes has no happy ending. But there is faith that medical science will develop better therapies, that researchers will inch closer to a cure, and that the human spirit will not succumb to this disease. And there is Garrett, laughing with his Friend for Life, knowing that he is not alone, bounding off the bus and heading for the Magic Kingdom, a little boy at play.

NOTES

BIBLIOGRAPHY

ACKNOWLEDGMENTS

INDEX

Notes

1. DIABETIC UTOPIA

13 *"carry more drama":* Gerald Alfred Wrenshall et al., *The Story of Insulin: Forty Years of Success against Diabetes* (Toronto: Max Reinhardt, 1962), 21.

14 *most common and costly:* David M. Cutler et al., "The Business Case for Diabetes Disease Management at Two Managed Care Organizations" (research paper, Harvard University, 2003), 1.

America's largest city, New York: New York Times, January 9, May 16, 2006.

"threats to human health": Nature, Vol. 414, December 13, 2001.

16 *"progressively obese population": Washington Post,* July 26, 2005.

17 *"not to overestimate numbers":* Author's interview.

18 *between 1988 and 1994:* Sharon H. Saydah et al., *JAMA,* January 21, 2004, Vol. 291. Data came from the National Health and Nutrition Examination Survey, conducted by the CDC.

leading research hospitals: Richard W. Grant et al., *Diabetes Care,* February 2005, 28: 337–442. Patients in poor control had hemoglobin A1c's of more than 9 percent.

"incompatible with human life": E-mail to author.

19 *only forty-two:* Jakob Larsen et al., *Diabetes,* August 2002, 51: 2637–41.

better diabetes management: The National Committee for Quality Assurance published its findings in its 2004 report, "State of Health Care Quality," and based its judgment on the Health Plan Employer Data and Information Set.

"killer can impose": E-mails to author.

22 *"lops off a leg":* Author's interview.

24 *"I don't think that they really understand":* Author's interview.

2. INSULIN'S POSTER GIRL

25 *"If father is elected":* Merlo J. Pusey, *Charles Evans Hughes* (New York: Macmillan, 1951), 330.

25 *"She walked out on air"*: Ibid., 608.

26 *"He began to starve her"*: "The Bittersweet Science," *New York Times,* March 16, 2003.

the "pissing evil": As described by the seventeenth-century English physician Thomas Willis.

patients withered away: All or parts of Aretaeus's text on diabetes have been translated into many books, including Chris Feudtner's *Bittersweet* (Chapel Hill: University of North Carolina Press, 2003), 4.

"honey or sugar": Ibid., 5.

27 *"occasionally analyzed his own"*: Elliott Joslin's speech to Eli Lilly & Company on September 23, 1946, the twenty-fifth anniversary of the discovery of insulin.

"Snow collected in Winter": *Diabetes Forecast,* December 2002.

28 *blood became more acidic*: This description of diabetic ketoacidosis comes largely from Michael Bliss, *The Discovery of Insulin* (Chicago: University of Chicago Press, 1984), 22.

"What sin has Pavy committed": Ibid., 24.

"are discharged dead": Joslin's speech to the Johns Hopkins Medical Society.

29 *"contains danger"*: *JAMA,* July 4, 1914, 939–43.

In 1917, a book: Lewis Webb Hill, M.D., and Rena S. Eckman, *The Starvation Treatment of Diabetes* (Boston: W. M. Leonard, 1917).

"Luxurious living": "Studies Concerning Diabetes," *JAMA,* July 4, 1914.

30 *"furnishes a few calories"*: *Starvation Treatment of Diabetes,* 9.

"There was a thrill": Joslin's speech, September 23, 1946.

31 *"obtained by confession"*: Bliss, 37.

"would have been unendurable": Ibid., 39.

"Inanition will undoubtedly": Joslin speech in 1918.

"undernutrition at the outset": Bliss, 36.

32 *"never take extras"*: From notes of the interview Hughes gave to Michael Bliss.

"more fun than camping": Letters from Elizabeth Evans Hughes and her mother are in the Banting Collection at the Fisher Library at the University of Toronto.

36 *"had not been out of bed"*: Bliss, 150–51.

"race of diabetics": Feudtner, 18.

"Now they make noise": "What I Teach My Diabetic Patients," *Diabetes,* January–February 1956.

37 *"salvation through insulin"*: Bliss, 161.

"exceeding great army": From Joslin's speech at the Lilly Research Laboratories in 1934.

38 *"scarcely able to walk"*: Ibid., 152.

"bread and potato": Ibid., 154.

41 *moved to Michigan*: Information about Elizabeth Evans Hughes's life in

Michigan comes from interviews with two of her children, Tony Denning and Tom Gossett.

43 *Elizabeth was keenly interested:* Information about Bliss's meeting with Hughes comes from his notes on file at the Fisher Library at the University of Toronto.

3. THE BURDEN OF CONTROL

45 *had treated 58,000:* According to Dr. Donald M. Barnett, who is writing Joslin's biography.

46 *"suffer and wither away":* Diabetes, March–April, 1956: 138.
"broke the Naunyn diet": Ibid.

47 *"completing his life's journey":* Joslin's quotes come from manuals, papers, and speeches in the library archive at the Joslin Diabetes Center in Boston.

49 *"injection is slight":* According to Dr. Barnett.

50 *Edward Tolstoi:* All quotes are from Edward Tolstoi, *Living With Diabetes* (New York, Crown, 1952).

53 *"history of science":* Author's interview.

54 *"fixation on causality":* Author's interview.
"hounded them": Author's interview.

55 *"control is important":* Deb Butterfield, *Showdown with Diabetes* (New York: W. W. Norton, 1999), 171.

4. THE DIABETES QUEEN

57 *"Fried chicken":* All Florene Linnen quotes are from interview with the author.

59 *diabetes and nutrition education:* According to data from the REACH 2010 coalition.

60 *"people of color sick":* Boston Globe, July 20, 2005.
Almost half of the adults: Type 2 diabetes in Starr County, Texas; *New York Times,* January 2, 2005.

61 *"colony of the sick":* New York Times, January 12, 2006.
"thrifty genes": The thrifty gene discussion, including the estimate from Achim Gutersohn, is from Francine R. Kaufman, *Diabesity* (New York: Bantam Dell, 2005).

62 *Pima Indians:* Information about the Pima Indians comes from Arthur Krosnick, "The Diabetes and Obesity Epidemic Among the Pima Indians," *New Jersey Medicine,* August 2000.

64 *Florene Linnen has a favorite:* All quotes from Georgetown from author interviews.

68 *"out of the question":* Health Scout News, October 28, 2002.

5. REWARDING FAILURE, PUNISHING EXCELLENCE

74 *"also an art":* All quotes from the Parkers are from author interviews.

76 *"middle-class patients":* Paul Starr, *The Social Transformation of American Medicine* (New York: Basic Books, 1982), 331.
"municipal, and state institutions": Charles E. Rosenberg, *The Care of Strangers* (New York: Basic Books, 1987), 345.

77 *Reimbursement was based:* Reimbursements for all but one procedure from *The New Yorker,* April 4, 2005. Reimbursement for the prosthetic from the *New York Times,* January 11, 2006.

78 *any other specialist:* Endocrinologist study, *Journal of Clinical Endocrinology & Metabolism,* Vol. 88, No. 5, April 2003.
C. Ronald Kahn: Quotes from author's interview.
pushed their limits: Kenneth M. Ludmerer, *Time to Heal* (New York: Oxford University Press, 1999), 384.

79 *"not a penalty":* Ibid., 385.
"tyranny of the urgent": Thomas Bodenheimer, *JAMA,* October 9, 2002.

80 *"a millionaire several times over":* Donald M. Barnett, "Elliott P. Joslin, M.D.: A Centennial Portrait" (Boston: Joslin Diabetes Center, 1998), 23.

80 *"post-operative infections":* *Washington Post,* July 24, 2005.

81 *"must help patients":* Quotes from Carolyn Swithers and Mary Whitlock come from author's interviews.

82 *40 percent of their profits:* "Healthcast Tactics: A Blueprint for the Future," a report from PricewaterhouseCoopers, 2002.
cost of bypass surgery: *Archives of Internal Medicine,* July 11, 2005.

6. YOU HAVE TO BE BRAVE, OR ELSE IT HURTS

93 *epidemic is focused:* "Type 1 Diabetes: Recent Developments," *British Medical Journal,* March 27, 2004.
by the age of twenty: Diabetes, 51: 3353–3361, 2002.

94 *"When I started, the average ages were ten to twelve":* says the program coordinator, Jo Ann Ahern. Author's interview.

95 *"broadies in the parking lot":* E-mail to author.
"take our shots together": Francine R. Kaufman, *Diabesity* (New York: Bantam Dell, 2005), 71–73.
"at 2 A.M.!": E-mail to author.

7. NEW LOWS

113 *"potato-chip fragile":* David Shenk, *The Forgetting* (New York: Anchor Books, 2001), 23.
estimated 2 to 4 percent: "Hypoglycemia in Diabetes," *Diabetes Care,* Vol. 26, No. 6, June 2003.

115 *"restless and agitated"*: Michael T. Kennedy, *A Brief History of Disease, Science & Medicine* (Cranston, R.I.: Writers' Collective, 2004), 407.

"would become rigid": Sylvia Nasar, *A Beautiful Mind* (New York: Touchstone, 2001), 292–94.

116 *Debra Hull:* The hypoglycemic experiences of Hull, Slobonik, Heffelfinger, Sparks, and Parent from author's interviews.

8. DR. BERNSTEIN'S SOLUTION

126 *"rape of the diabetic"*: Richard K. Bernstein, *Dr. Bernstein's Diabetes Solution* (Boston: Little, Brown, 1997), 112.

129 *"live on high carbs"*: All quotes are from interviews with the author.

9. HIGH-TECH TRADEOFFS

137 *to near 100,000:* Byetta sales data from the *New York Times,* March 2, 2006.

139 *"This is nuts"*: All quotes from author's interviews.

10. PUSHING BACK THE HORIZON

152 *Gary Kleiman:* Information about Gary Kleiman, including quotes, come from the author's interviews or *No Time to Lose,* by Gary Kleiman and Sanford Dody (New York: William Morrow, 1983).

154 *"discovery of insulin": Diabetes,* January–February 1956: 68.

155 *"stabbed me right"*: All quotes from Lee Ducat and Carol Lurie from author's interviews.

157 *"their dimes directly"*: David M. Oshinsky, *Polio: An American Story* (Oxford, U.K.: Oxford University Press, 2005), 54.

160 *"islet transplantation": Scientific American,* July 1995, 51.

161 *"sleeping because he's tired"*: Marge Kleiman's quotations from author's interview.

164 *"absolute miracle": Science Magazine,* June 20, 2003.

"going bald": Chris McAliley's quotations from author's interview.

167 *"inexorable" loss:* James Shapiro's comments at the Canadian Diabetes Association in Edmonton in October 2005, as reported by *Diabetes Close Up.*

169 *"desperate straits"*: Ibid.

11. THE MAGICAL BETA CELL

171 *"how life develops"*: Unless otherwise indicated, quotes from Doug Melton and Gail O'Keefe are from the author's interviews.

"Honey": "Son's Disease Propels a Stem Cell Pioneer," *Boston Globe,* March 20, 2005.

174 *"than a potato":* Michael Kinsley, "Dance of the Stem Cells," *Washington Post,* August 15, 2004.

"precious time and effort": "Science Test: Biggest Struggles in Stem-Cell Fight May Be in the Lab," *Wall Street Journal,* August 12, 2004.

176 *behaved like "popcorn":* Ibid.

178 *"four dishes for dinner":* Ibid.

180 *"work that we envisioned":* "Harvard Teams Want OK to Clone," *Boston Globe,* October 13, 2004.

"I believe that Harvard": "Son's Disease Propels a Stem Cell Pioneer," *Boston Globe,* March 20, 2005.

181 *"a puzzle":* "Towards a Diabetes Cure," *Harvard Science Review,* 2004.

182 *"begins at fertilization":* David Ewing Duncan, *The Geneticist Who Played Hoops with My DNA . . . And Other Masterminds from the Frontiers of Biotech,* excerpt from *Discover,* June 2005.

"Like many scientists": Ibid.

"unfettered creativity": "The President and the Scientists," *The New Yorker,* March 13, 2006.

12. THE TRIALS OF A MAVERICK SCIENTIST

184 *"goofy niece":* All quotes from Joseph Avruch from the author's interview.

"green thumb": Interview with Denise Faustman. All quotes from Faustman are from the author's interviews unless otherwise indicated.

186 *"scientific dogma":* Biography published on NIH's Web site: "Celebrating America's Women Physicians."

188 *"academic investigator":* Author's interview.

189 *"make the breakthroughs":* Author's interview.

190 *"superb success":* "Fostering Innovation and Discovery in Biomedical Research," *JAMA,* September 21, 2005.

192 *"invisible tissue":* "Transplant Trick," *Straits Times,* July 2, 1991.

"it was bad": All quotes from Lee Iacocca from the author's interview unless otherwise indicated.

193 *"You think I have it bad?":* Lee Iacocca with William Novak, *Iacocca: An Autobiography* (New York: Bantam, 1984), 288.

200 *"autoimmune response":* Camillo Ricordi's speech at the Tenth World Congress of the International Pancreas Islet Transplantation Association, in Geneva, Switzerland, in May of 2005, as reported by Close Concerns' annual report.

209 *"respect for Dr. Faustman":* National Journal, January 21, 2005.

"one of the most gifted": Francine R. Kaufman, *Diabesity* (New York: Bantam Dell, 2005), 39.

210 *"tough to get any work done":* Author's interview.

213 *"public up until":* Author's interview.

214 *"proxy for our ability"*: "Why Did the JDRF Try to Discredit Researcher?" *Diabetes Health,* June 2005.
"created a cult following": "Science-Charitable Choices," *National Journal,* January 21, 2005.

13. THE PRICE OF SURVIVAL

220 *success of HealthPartners:* Information on HealthPartners from "The Business Case for Diabetes Disease Management at Two Managed Care Organizations," by Nancy Dean Beaulieu, David M. Cutler, Katherine E. Ho, Common Wealth Fund, April 2003.

222 *"paid for chronic care"*: "Improving Primary Care for Patients with Chronic Illness," *JAMA,* October, 16, 2002.
Puget Sound: Ibid.
study by Kaiser: Ibid.

223 *One study at three clinics in California: Diabetes Care,* January 2004. Jovanovic quotes from author interview.

226 *"integrated health information"*: *Washington Monthly,* January/February 2005.
1,285 diabetics: Annals of Internal Medicine, August 17, 2004.

227 *"last ten years"*: *Washington Post,* August 22, 2005.

228 *"these look like"*: Author's interview.

229 *"I damn well wasn't"*: Author's interview.

232 *"I predict it will"*: Author's interview.

233 *"the defect that distinguishes"*: E-mail to author.

237 *innate system:* Discussion of the innate immune system also drawn from "The Future of Organ and Tissue Transplantation," *JAMA,* September 15, 1999.

238 *"often so efficient"*: John M. Barry, *The Great Influenza* (New York: Penguin, 2004), 247.

14. SURVIVOR TALES

242 *"cured me tomorrow"*: Spiro's quotes come from author's interview.

243 *"given the freedom"*: Fisher's quotes come from a college essay and author interview.

246 *"Leave it alone"*: *Washington Post,* May 16, 1997.

247 *this one for life:* All prison information comes from Bennett's letters to the author.

249 *belongs to Eva Saxl:* In addition to the author's interview with Eva Saxl, information came from three articles in *Diabetes Forecast:* July 1991, May–June 1984, and January–February 1979; and an article in *Diabetes Interview,* January 2002. Other information came from Sáxl's letters to Richard Bernstein and an oral history of her life taken by Zena M. M. Colterjohn.

Bibliography

Angell, Marcia. *The Truth about Drug Companies*. New York: Random House, 2004.

Barry, John M. *The Great Influenza*. New York: Penguin, 2004.

Bernstein, Richard K. *Dr. Bernstein's Diabetes Solution*. Boston: Little, Brown, 1997.

Bliss, Michael. *Banting*. Toronto: University of Toronto Press, 1984.

——— . *The Discovery of Insulin*. Chicago: University of Chicago Press, 1982.

Butterfield, Deb. *Showdown with Diabetes*. New York: W. W. Norton, 1999.

Cutler, David. *Your Money or Your Life*. Oxford, U.K.: Oxford University Press, 2004.

Connors, Mary Frances. *Sweet Blood & Fury*. Sunnyvale, Calif.: Far Western Graphics, 2000.

Cunningham, Robert, III, and Robert M. Cunningham, Jr. *The Blues*. DeKalb, Ill.: Northern Illinois University Press, 1997.

Dawson, Leslie Y. *How to Save up to $3,000 a Year on Your Diabetes Costs*. American Diabetes Association, 2004.

Dominick, Andie. *Needles*. New York: Touchstone, 1998.

Dubos, René. *Mirage of Health*. New Brunswick, N.J.: Rutgers University Press, 1987.

Feudtner, Chris. *Bittersweet*. Chapel Hill: University of North Carolina Press, 2003.

Fox, Renee C., and Judith P. Swazey. *The Courage to Fail*. New Brunswick, N.J.: Transaction, 2002.

Gingrich, Newt, with Dana Pavey and Anne Woodbury. *Saving Lives &*

Saving Money. Washington, D.C.: Alexis de Tocqueville Institution, 2003.

Groopman, Jerome. *The Anatomy of Hope.* New York: Random House, 2004.

Hall, Stephen. *Invisible Frontiers.* Oxford, U.K.: Oxford University Press, 1987.

Iacocca, Lee, with William Novak. *Iacocca.* New York: Bantam, 1984.

Ingelfinger, Franz J., Arnold S. Relman, and Maxwell Finland, eds. *Controversy in Internal Medicine.* Philadelphia: W. B. Saunders, 1966.

Ingelfinger, Franz J., Richard V. Ebert, Maxwell Finland, and Arnold S. Relman, eds. *Controversy in Internal Medicine II.* Philadelphia: W. B. Saunders, 1974.

Institute of Medicine. *Crossing the Quality Chasm.* Washington, D.C.: National Academy Press, 2001.

Johnson, Nicole. *Living with Diabetes.* Washington, D.C.: LifeLine, 2001.

Kaufman, Francine R. *Diabesity.* New York: Bantam, 2005.

Kennedy, Michael T. *A Brief History of Disease, Science & Medicine.* Cranston, R.I.: Writers' Collective, 2004.

Kleiman, Gary, and Sanford Dody. *No Time to Lose.* New York: William Morrow, 1983.

Lax, Eric. *The Mold in Dr. Florey's Coat.* New York: Henry Holt, 2004.

Lodewick, Peter A. *A Diabetic Doctor Looks at Diabetes.* Cambridge, Mass.: RMI Corporation, 1982.

Ludmerer, Kenneth M. *Time to Heal.* Oxford, U.K.: Oxford University Press, 1999.

Madison, James H. *Eli Lilly.* Indianapolis: Indiana Historical Society, 1989.

Marmor, Theodore R. *Understanding Health Care Reform.* New Haven, Conn.: Yale University Press, 1994.

Nasar, Sylvia. *A Beautiful Mind.* New York: Touchstone, 2001.

Nathan, David M. *Diabetes.* New York: Times Books, 1997.

Oshinsky, David M. *Polio.* Oxford, U.K.: Oxford University Press, 2005.

Papaspyros, N. S. *The History of Diabetes Mellitus.* Stuttgart, Germany: Georg Thieme, 1964.

Parson, Ann B. *The Proteus Effect.* Washington D.C.: Joseph Henry Press, 2004.

Porter, Roy. *The Greatest Benefit to Mankind.* New York: W. W. Norton, 1997.

Pusey, Merlo J. *Charles Evans Hughes, Volumes One and Two.* New York: Macmillan, 1951.

Rapaport, Wendy Satin. *When Diabetes Hits Home.* American Diabetes Association, 1998.

Reflections on a Life with Diabetes, edited by Diane M. Parker and Ruth Mark. College Station, Tex.: Virtualbookworm.com, 2004.

Roney, Lisa. *Sweet Invisible Body.* New York: Henry Holt, 1999.

Rosenberg, Charles E. *The Care of Strangers.* New York: Basic Books, 1987.

Rosenthal, Helen. *Diabetic Care in Pictures.* Philadelphia: J. B. Lippincott, 1946.

Rubin, Richard R., June Bierman, and Barbara Toohey. *Psyching Out Diabetes.* Los Angeles: Lowell House, 1997.

Sanders, Lee J. *The Philatelic History of Diabetes.* American Diabetes Association, 2001.

Shell, Ellen Ruppel. *The Hungry Gene.* New York: Atlantic Monthly Press, 2002.

Shenk, David. *The Forgetting.* New York: First Anchor Books, 2001.

Shilts, Randy. *And the Band Played On.* New York: Quality Paperback Book Club, 1987.

Solomon, Andrew. *The Noonday Demon.* New York: Touchstone, 2001.

Starr, Douglas. *Blood.* New York: Perennial, 2002.

Starr, Paul. *The Social Transformation of American Medicine.* New York: Basic Books, 1982.

Stevens, Rosemary. *American Medicine and the Public Interest.* New Haven, Conn.: Yale University Press, 1971.

The Journey and the Dream. American Diabetes Association, 1990.

Thomas, Lewis. *The Lives of a Cell.* New York: Bantam, 1974.

Tolstoi, Edward. *Living with Diabetes.* New York: Crown, 1952.

Tompkins, Walker A. *Continuing Quest.* Santa Barbara, Calif.: Sansum Medical Research Foundation, 1977.

Twist, Michael. *Highs & Lows.* Toronto: Insomniac Press, 2001.

Vernon, Mary C., and Jacqueline A. Eberstein. *Atkins Diabetes Revolution.* New York: William Morrow, 2004.

Weiner, Jonathan. *His Brother's Keeper.* New York: HarperCollins, 2004.

Wrenshall, G. A., G. Hetenyi, and W. R. Feasby. *The Story of Insulin.* Toronto: Max Reinhardt, 1962.

Acknowledgments

If one of the themes of this book is the special role that parents play in this disease, it's fitting that I extend my first acknowledgment to my mother.

Gloria Hirsch did not have diabetes, but as both a parent and a volunteer, she devoted much of her adult life to the cause. For almost three decades, she chaired a bikeathon in St. Louis that raised money for the ADA. While she had no professional training in marketing, she devised creative promotional ideas. One year, a diabetic monkey served as the mascot, sitting on a bike for a publicity shot that appeared on local newscasts and in newspapers. To put together a free trip to Walt Disney World, she asked friends to donate their frequent flier miles. She pleaded even for the smallest prizes — stuffed animals, baseball gloves, cosmetics. She wanted as many riders as possible, mostly children and teenagers, to walk away as winners.

When my mom began this work in the middle 1970s, raising $35,000 to $50,000 from one event was considered serious money, but all of her bikeathons combined probably didn't raise much more than a million dollars — a bad night for some JDRF galas. My mom wasn't slick or visionary, but she had a wonderful, naïve faith in the proposition that regular people working hard, working together, could defeat this disease.

After she was diagnosed with cancer in 2001, the ADA honored her at a dinner, announcing that the camp where her children had once worked had been renamed the Gloria Hirsch Camp for Diabetic Children. She died the following year, and at her funeral service, the rabbi discussed her work in diabetes, noting that my father had been her indispensable partner in all her efforts. He talked as well about my brother's contributions; our sister, Lynn, had also been a counselor for many summers at the camp. "When they finally find a cure for this disease," the rabbi said, "this family will have played a significant role in that achievement."

Everyone in my family had indeed done a lot. But near as I could tell, I had done very little, other than survive. So it was at that moment that I decided I would write a book about diabetes. I didn't know when, or what it would say, but I hope this book is seen as part of my family's contribution to the cause.

I actually didn't begin this project until 2004, and my brother was my catalyst. Irl told me about a ceremony in which Eli Lilly would be recognizing patients who'd been taking insulin for more than fifty years. These "insulin pioneers" appealed to me as a starting point for a book, and in writing about the likes of Elizabeth Evans Hughes, Eva Saxl, and Robert Spiro, I wanted to pay homage to all those who, despite primitive medical care, persevered with courage, resilience, and grace.

If you're going to write a book about diabetes, you could do far worse than have Irl Hirsch as your brother. In a field of many rivalries, he seems to command the respect of everyone. He can barely walk through a lobby at a diabetes convention without getting mobbed. His name opened many doors for me in my own research, and I'm grateful for his encouragement. I'm proud to dedicate the book to him and to our sister, Lynn Friedman, who's always looked out for me and whose professional and personal achievements are also exemplars for the family. I appreciate as well the support of my siblings' own families — Ruth and Barbara Hirsch and Howard, Sam, and Max Friedman. I'm grateful to my father, Ed Hirsch, and hope that Garrett will hold me in the same esteem that I hold my dad. I also thank Dolly Newport for enriching all of our lives.

Sheryl's family deserves special mention for its love and support: Aileen, Leslie, and Harris Phillips, Marlane and Ben Pinkowitz, and Lori and Rob Cohen. A special thanks to, and blessing for, Randy Phillips.

Early in my research, I placed a notice in the "Making Friends" section of *Diabetes Forecast,* where readers briefly describe themselves in hopes of finding a pen pal. I wrote that I was writing a book about diabetes and would like to hear unusual or telling stories about readers' experiences — the strangest place they've ever given an injection, for example, or their worst hypoglycemic incident. I got more than 700 letters and e-mails (including multiple e-mails from the same people). In many ways, these responses form the spine of this book, for they conveyed a diabetic world — of anger, guilt, and sorrow; of humor, hope, and faith — that I had never appreciated. Some letters were twenty pages long. Other correspondents sent me diabetic manuals from the 1930s or '40s that had been used by long-deceased relatives. One thing was clear: they wanted their story told. I tried to thank as many as I could through letters and e-mails, but I couldn't keep up. I am, however, in debt to every correspondent. The list is too long to thank by name, and to cite any person may be unfair to the rest, but I will mention two. Karen Katsamore wrote with unusual eloquence and candor about her many experiences, including that of her stillborn daughter. Debra Hull became my kindred spirit as she shared the fears and frustrations of the disease, and we forged a bond that will last long after the publication of this book.

Most of the chapters are driven by specific characters, and I would like to thank those individuals who allowed me to tell their story. Florene Linnen opened the doors to Georgetown County, South Carolina, so that I could understand the epidemic in minority communities. Steven and Sonny Parker laid bare the painful demise of their medical practice with insight and honesty. Richard Bernstein and Anna Smith allowed me to observe and describe their meeting in Bernstein's office. If Gary Kleiman was an ideal candidate for an islet cell transplant, he was an even better subject to interview — thoughtful, friendly, candid, and funny. Doug Melton is uncommonly gifted in describing the complexities of embryonic stem cells in ways that laypeople can understand, and Gail O'Keefe shared the burdens of having children with diabetes and a husband dedicated to curing them. Denise Faustman spent many hours with me in her lab, talking about her research and the controversy that's followed her. I cannot imagine three people more different than Cathy Fisher, Eugene Bennett, and Robert Spiro (a college student, a prisoner, and a scientist), but I appreciate that they all shared their exceptional stories. And if I had to entrust my son's life to one researcher, I would choose

David Harlan — not because he's smart, truthful, and hardworking, but because his medical views are informed by history, literature, and philosophy. He would treat the soul as well as the body.

My thanks to Saundra Ketner, the librarian at the Alexander Marble Library at the Joslin Center. I spent many hours in the Special Collections Room and would often bump into Donald Barnett, a former physician at the clinic who is working on a biography of Elliott Joslin. Barnett kindly shared with me his insights and some of his research. Lisa Bayne, the archivist at Eli Lilly, graciously helped me sort through the company's insulin materials. Michael Bliss, in addition to writing the definitive account of insulin's discovery, gave me a superb historical framework for diabetes, and Kenneth Ludmerer, a medical historian at Washington University, provided me with a similar foundation for understanding America's health care system.

The following is a partial list of others who helped my research: Audrey Adreon, William Ahearn, Jo Ann Ahern, Ann Albright, Jeff Asher, Ronald Arky, Joseph Avruch, Nicole Johnson Baker, Dana Ball, Michael Brownlee, Douglas Burger, Maria Buse, Marilyn Cattanach, Joan Chamberlain, Peter Cleary, Zena Colterjohn, John Colwell, Sonia Cooper, Lee Ducat, Daniel Duick, Jay Dunigan, Robert Eckel, Linda "Freddi" Fredrickson, John Galgani, Robert Galvin, Martin Goldberg, David Goldstein, Claudia Graham, Richard Guthrie, Debbie Hinnen, Melissa Hogan-Watts, John Holcombe, Sherman Holvey, Lee Iacocca, Giuseppina Imperatore, Richard Insel, Carolyn Jenkins, Martin Jensen, Emily Jones, Jeannette Jordan, Lois Jovanovic, Richard Kahn, Ronald Kahn, Jane Kelly, Glenn Lampaert, Francine Kaufman, Pamela King, Arthur Krosnick, Karmeen Kulkarni, Russell LaMontagne, Carol Lurie, Paul Madden, John Mastrototaro, Chris McAliley, Daniel Mintz, Marge Kleiman Mintz, David Nathan, Jeremy Nobel, Joy Pape, Alisha Perez, Margery Perry, Ken Quickel, Larry Raff, Wendy Satin Rapoport, Paige Reddan, Tim Reid, Bertha Rice, Susan Root, Aldo Rossini, Louis Sandler, David Scharp, David Serreze, Kantha Shelke, Jay Skyler, Kathleen Slovacek, Christina Smith, Richard Smith, Larry Soler, Brien Stafford, Caroline Stevens, Carolyn Swithers, Ellen Ullman, Peter Van Etten, James Warram, Hope Warshaw, Neil White, Mary Whitlock, Howard Wolpert, Anne Woodbury, and Mona Zawaideh.

A special thanks to Connie Hubbell, a loyal friend and excellent colleague. Kelly Close understands the intersection of diabetes and business as well as anyone, and she and her colleague Erin Kane gave me valuable feedback on the manuscript. Carol Verderese, an excellent medical writer, did as well.

I'm grateful to the many people who helped us after Garrett's diagnosis. That includes Dr. Jamie Wood and Dr. David Breault for their tender care in the hospital; to Cindy Pasquarello, our nurse at Joslin, for her warmth and guidance; to the JDRF volunteer Lois Hearn, who met us after we returned from the hospital; and to Jeff Hitchcock, who's done so much for so many parents, myself included. We're fortunate to have met wonderful new friends, Elizabeth and Keith Wexelblatt, whose six-year-old son, Jake, is flourishing on the pump and will someday play for, run, or own the Boston Red Sox. We're in debt to our friends and neighbors, Laura and John Wiesman, who were always there when we needed them; to Donna McKeown, a longstanding friend of Sheryl's who loves our children like her own; and to Rabbi Jay Perlman, for his counsel, compassion, and friendship.

We were truly blessed that Garrett attended Carter Center for Children in Needham, to which this book is also dedicated. The directors — Barbara Carr and Peggy McDonald — assured us that Garrett would always be welcomed with open arms, and the teachers embraced him with tender vigilance and boundless affection. Garrett will never be in an environment, other than his own home, as secure and loving as Carter.

To Philip Channen, Marty Goldberg, Gary Kaufman, Chris Martin, Michael Seidman, Alan Shapiro, and Bruce Tribush — long live the Needham Poker Hacks!

I'm grateful, as always, to my agent and friend Todd Shuster, who encouraged me to write this book from the outset. In our first conversation, I tried to explain the life of a diabetic. "It's a constant struggle," I said, "to always stay one step ahead of it, to prevent complications and death, to cheat destiny."

He stopped me in my tracks. "That's your title!" he said. "Cheating Destiny." When your agent knows your title before you do, you know you're really "on the same page."

Finally, I appreciate Houghton Mifflin's continued support, specifically that of Eamon Dolan. In this case, he helped me turn a deeply felt but amorphous idea into a coherent narrative, and he encouraged a stronger, more personal voice. As usual, because of Eamon, I'm a better writer now than when I began this book. My thanks as well to Anne Seiwerath for valuable assistance along the way. And my manuscript editor, Luise Erdmann, improved this book through her assault on unnecessary digressions, gratuitous syllables, and pretentious language. If every editor changed "ophthalmologic and podiatry services" to "eye and foot care," the world would be a better place.

This is the point where I'm supposed to thank my wife and children, but that seems rather hollow. I usually hope that Sheryl, Amanda, and Garrett will be patient and supportive when I'm writing a manuscript, and they always are. Little did I know that, in this case, they would all be central characters. They did for the book what they do for me: they made it whole.

May 2006

Index

James S. Hirsch, a former reporter for the *New York Times* and the *Wall Street Journal,* is the author of *Hurricane: The Miraculous Journey of Rubin Carter, Riot and Remembrance: The Tulsa Race War and Its Legacy,* and *Two Souls Indivisible: The Friendship That Saved Two POWs in Vietnam.* He is also a principal in Close Concerns, a consultancy and publishing company that specializes in diabetes. He lives in the Boston area with his wife, Sheryl, and their children, Amanda and Garrett.

HURRICANE

THE MIRACULOUS JOURNEY
OF RUBIN CARTER

A New York Times Bestseller

"Hirsch writes vividly and tells the entire story with economy and grace."
— *Boston Globe*

In 1967, the black boxer Rubin "Hurricane" Carter and a young acquaintance were wrongly convicted of triple murder by an all-white jury in Paterson, New Jersey. As Carter gradually amassed convincing evidence of his innocence, he would achieve a freedom more profound than any that could be granted by a legal authority.

ISBN 978-0-618-08728-0

RIOT AND REMEMBRANCE

AMERICA'S WORST RACE RIOT
AND ITS LEGACY

"An illuminating and brilliant discussion of history, memory, and forgetting."
— *Washington Post Book World*

On May 30, 1921 in Tulsa, Oklahoma, a misunderstanding between a white elevator operator and a black delivery boy escalated into the worst race riot in U.S. history. In this compelling and deeply human account, Hirsch investigates how it erupted, how it was covered up, and how the survivors and their descendants are fighting for belated justice.

ISBN 978-0-618-34076-7

TWO SOULS INDIVISIBLE

THE FRIENDSHIP THAT SAVED
TWO POWs IN VIETNAM

"A moving story of two men whose courage, sense of duty, and love proved greater than the depravity of their captors." — **Senator John McCain**

A powerfully reported, unforgettable true story, *Two Souls Indivisible* stirringly recounts the story of Fred Cherry and Porter Halyburton, who first met in their shared cell in a brutal POW camp in Vietnam. Their intense connection would sustain both men through the war and throughout their lives.

ISBN 978-0-618-56210-7

CHEATING DESTINY

LIVING WITH DIABETES

"The book that people who care about diabetes have been waiting for . . . Hirsch persuasively illustrates an epidemic that is at odds with modern society at almost every level."
— *Washington Post*

In an engaging blend of history, reportage, advocacy, and memoir, Hirsch — who has lived with type 1 diabetes for more than three decades — crafts an incisive, surprising portrayal of the science behind the disorder and its skyrocketing impact on our economy, our society, and our families.

ISBN 978-0-618-91899-7